# GUIDE TO FOOD LAWS AND REGULATIONS

# GUIDE TO FOOD LAWS AND REGULATIONS

Patricia A. Curtis

**Patricia A. Curtis** is a professor and the director of the Poultry Products Safety and Quality Peaks of Excellence Program, Department of Poultry Science, Auburn University, and approved lead instructor and accredited course provider for the International HACCP Alliance.



Blackwell Publishing Professional
2121 State Avenue, Ames, Iowa 50014, USA

Orders: 1-800-862-6657
Office: 1-515-292-0140
Fax: 1-515-292-3348
Web site: www.blackwellprofessional.com

Blackwell Publishing Ltd
9600 Garsington Road, Oxford OX4 2DQ, UK
Tel.: +44 (0)1865 776868

Blackwell Publishing Asia
550 Swanston Street, Carlton, Victoria 3053, Australia
Tel.: +61 (0)3 8359 1011



First edition, 2005

Library of Congress Cataloging-in-Publication Data

Curtis, Patricia A.
Guide to food laws and regulations / Patricia A. Curtis.—1st ed.
p. cm.
Includes bibliographical references and index.
ISBN-13: 978-0-8138-1946-4 (alk. paper)
ISBN-10: 0-8138-1946-6 (alk. paper)
1. 1. Food law and legislation—United States. I. Title.

KF3870.C87 2005
344.7304′232—dc22

2004030347

The last digit is the print number: 9 8 7 6 5 4 3 2

# Contents

# Contributors

Brooke Caudill
1387 Hampton Dr.
Auburn, AL 36830;
Phone: 334-887-6198
bablingb@yahoo.com
Chapter 10

Muhammad M. Chaudry, Executive Director
Islamic Food and Nutrition Council
5901 N. Cicero Ave., Suite 309
Chicago, IL 60646
Phone: 773-283-3708
Fax: 773-283-3973
mchaudry@ifanca.org
Chapter 9

Patricia Curtis
260 Lem Morrison Dr.
Poultry Science Department
Auburn University
Auburn, AL 36849
Phone: 334-844-2679
Fax: 334-844-2641
Pat_Curtis@Auburn.edu
Chapters 1, 2, 3, 4, 5, 7, 8, 10

Wendy Dunlap
430 Rolling Pines Lane
Duncan, SC 29334
Phone: 864-433-3332
Fax: 864-433-3146
wendy.j.dunlap@sealedair.com
Chapters 1, 2, 3, 4, 5

Theodore A. (Ted) Feitshans
North Carolina State University
Dept. of Agric. & Resource Economics
Room 3340 Nelson Hall
Campus Box 8109
Raleigh, NC 27695-8109
Phone: 919-515-5195
Fax: 919-515-6268
ted_feitshans@ncsu.edu
Chapter 6

Carrie E. Regenstein
Associate Chief Information Officer
Associate Director, Division of Information Technology
University of Wisconsin-Madison
Madison, WI 53706
Chapter 9

Joe M. Regenstein
Professor of Food Science
Cornell Kosher Food Initiative
Department of Food Science
Stocking Hall
Cornell University
Ithaca, NY 14853-7201
Phone 607-255-2109
Fax: 607-257-2871
jmr9@cornell.edu
Chapter 9

# Preface

Food laws and regulations change frequently. The best place to find the most current law or regulation on a specific subject is from the government website relevant to the topic. Therefore, the content of this book is composed primarily from excerpts of government websites. The focus of the book is not to provide you with the most current law or regulation, but to provide you with the knowledge of where to find the most current information on the subject you seek.

# Acknowledgments

Writing a book while working full time takes away all spare time and leaves no time for family responsibilities. Therefore, I would like to thank my family—Ben, Heather, Dylan, Myles, and Detrych—for their patience and support. Many thanks to my coworkers and graduate students—Marcia, Lesli, Regina, Vanessa, Abby, Brooke, and Alexis—who took on numerous additional tasks in order to free up my time for writing. And last, but not least, I would like to thank my coauthors without whose assistance this book would not be possible.

# GUIDE TO FOOD LAWS AND REGULATIONS

CHAPTER 1

# An Introduction to Laws and Regulations

Patricia Curtis, Auburn University
Wendy Dunlap

## Introduction

The American democracy is based on the following six essential principles.

1. The majority rules
2. Protection of political rights of minorities
3. Citizens agree to be ruled by a system of law
4. Free exchange of ideas and opinions
5. Equality of all citizens
6. Government exists to serve the people

In the United States, the combination of federal, state, and local laws, bodies, and agencies are responsible for carrying out operations. This combined group ensures that the people are the source of the government's authority by electing representatives to serve in the government in all levels and provides for checks and balances by sharing power between different levels of government.

The president and vice president are the only public officials elected by all the citizens of the United States. Each serve a four-year term and are eligible for an additional four-year term.

Each president adds personal touches to the Oval Office during the term of occupancy. A brief biography of each of the previous presidents can be found at http://www.whitehouse.gov/kids/presidents/index.html.

## Sources of American Law

There are four sources of American law: the Constitution, statutory law, common law, and equity.

I. Constitution. The Constitution is the supreme law of the United States. It describes what powers the government has, as well as what rights U.S. citizens have. All other laws must comply with the Constitution. It has six basic principles.

1. Popular sovereignty. The people have the power to govern. Likewise, the people must entrust this power to their government. People elect their congressmen, who make the laws that govern them.
2. Separation of powers. The U.S. government is divided into three branches:

    - The legislative branch, which is responsible for law making.
    - The judicial branch, which is responsible for law interpreting.
    - The executive branch, which is responsible for law enforcement.

    Each of these branches has its own responsibilities, constituencies, and organization.
3. Checks and balances. No branch of the government can act completely on its own. Each branch has some control over the other two branches. To make a law, Congress (legislative branch) must get an okay from the president (executive branch), except in special cases.
4. Federalism. Under the federalist form of government, federal, state, and local governments exist and have their own powers. This separation of government power helps prevent abuse of that power. In general, state laws deal with matters that are contained within the state's borders. The state laws must be as stringent as the federal laws and must comply with the Constitution. For example, the North Carolina Department of Agriculture is allowed to regulate food that is produced and sold within the state boundaries.
5. Supremacy of national laws. When a federal and state law contradict, the federal law will be upheld.
6. Civilian control of government. Limits are placed on military power by putting control of the military into civilian hands. The president (a civilian) is the commander in chief, and only Congress can approve war and defense spending.

II. Statutory law. Statutory law is written law that is passed by legislatures. Congress, state legislatures, and local governments all enact statutes. While Constitutional law is broad and leaves room for interpretation, statutory law is generally more detailed and precise. Regulations passed by agencies are even more specific than statutes.

III. Common law. The laws based on previous court ruling are called common, or case, law. This system dates back to eleventh century England where judges contemplating a case would refer to previous case rulings. After seeing what was common, he would then make his decision. When common law is in conflict with statutory law, the statutory law is upheld.

IV. Equity. Equity cases deal with the fairness or justice of a situation. Judges decide the issues, and a jury is not present. The judge often orders injunctions to prevent the unfair act from happening again.

## Public and Private Law

There are two branches of law in America (not to be confused with the three branches of government): private law and public law.

Private laws deal mainly with disputes between individuals, businesses, or other organizations. The outcome of these disputes is usually a fine or award of money as opposed to a jail sentence. Private law encompasses property, contracts, family relations, and torts.

Public laws deal with the relationship between the government and its citizens. The four categories of public law are constitutional law, international law, criminal law, and administrative law. Administrative law encompasses the rules and regulations that governmental agencies make. The bulk of rules and regulations that control food quality and safety fall into this category.

## The Legislative Branch

The legislative branch is a bicameral system, which means that it is composed of two houses, the Senate and the House of Representatives, as outlined in the United States Constitution.

The U.S. Senate is made up of 100 members, two elected from each state. The U.S. House of Representatives is composed of 435 members elected every two years from among the 50 states, apportioned to their total populations.

These two houses together form the U.S. Congress and are mainly responsible for passing statutory, also known as legislative, laws.

The Constitution gives specific powers to Congress. These are

- to levy and collect taxes;
- to borrow money for the public treasury;
- to make rules and regulations governing commerce among the states and with foreign countries;
- to make uniform rules for the naturalization of foreign citizens;
- to coin money, state its value, and provide for the punishment of counterfeiters;
- to set the standards for weights and measures;
- to establish bankruptcy laws for the country as a whole;
- to establish post offices and post roads;
- to issue patents and copyrights;
- to set up a system of federal courts;
- to punish piracy;
- to declare war;
- to raise and support armies;
- to provide for a navy;
- to call out the militia to enforce federal laws, suppress lawlessness or repel invasions by foreign powers;
- to make all laws for the District of Columbia; and
- to make all laws necessary to enforce the Constitution.

The vice president is the president of the Senate. He or she has a vote only in the case of a tie. A president *pro tempore* is chosen by the Senate to preside when the vice president is absent. The House of Representatives chooses its Speaker of the House.

Congress also has the power to investigate. This includes investigating the need for new legislation and the effectiveness of existing legislation and evaluating the qualifications and performance of members of the executive and judicial branches. The House of Representatives is responsible for conducting impeachment proceedings, and the Senate is responsible for impeachment trials.

## Judicial Branch

The judicial branch consists of the federal court system, with the Supreme Court being the top entity. It is the judicial branch's responsibility to interpret the intent of laws and to settle disputes.

Its jurisdiction includes cases involving the Constitution, controversies when the U.S. government is a party, and controversies between states or their citizens. In food law, this often includes disputes between an enforcing agency and a food company.

The U.S. Congress has the power to create and abolish federal courts. It cannot abolish the Supreme Court, however. The president appoints U.S. judges, who must then be confirmed by the Senate.

## Federal Court System

### The Supreme Court

The Supreme Court is the highest court in the federal court system and was created by the Constitution. Its decisions cannot be overturned. About 10 percent of its cases get to the Supreme Court by appeals. The rest of the cases get there through writs of certiorari (sir-shee-uh-RARE-ee). This is an order telling a lower court to send its records of a case to the Supreme Court for review. Writs of certiorari are issued for cases involving a serious constitutional issue or an error in the lower court. The Supreme Court consists of one chief justice and eight associates.

### Courts of Appeals

If there is a question regarding the fairness of a trial, a case can be appealed to the court of appeals. Here, the case is reviewed by a panel of judges who determine if the district court decision was correct. If they need clarification on a point, they can ask to hear oral statements.

These courts are also known as circuit courts, and they are arranged according to geography, with 12 circuits in all.

### U.S. District Courts

There are 94 federal district courts. These courts have original jurisdiction over both criminal and civil federal laws. In criminal cases, a grand jury decides if there is enough evidence to try the case. A trial jury then determines if the person is guilty. In civil cases, a trial jury can be used or can be waived if both parties agree. In these cases, a federal judge makes the final decision.

### Special Courts

Congress has set up special courts to deal with specific problems. One of these courts is the Court of Customs and Patent Appeals. This court handles issues involving the U.S. Patent Office.

## Executive Branch

The executive branch includes the president, his cabinet, the cabinet departments, and independent agencies. The president has many powers. These include the power to

- introduce legislation to Congress,
- veto legislation,
- appoint federal judges,
- grant full and conditional pardons,
- call the National Guard into service,
- appoint ambassadors, ministers, and consuls to aid in foreign relations, and
- appoint heads of the executive departments and independent agencies.

The heads of the executive departments make up the president's cabinet. These cabinet departments, along with independent agencies also in the executive branch, are responsible for enforcing laws passed by the legislative branch. The responsibility of food safety and quality is spread out among four cabinet-level departments and two independent agencies.

## Sources of Legislation

For the purpose of this discussion, the word *law* refers to statutory law or laws passed by Congress. As mentioned earlier, the U.S. Congress is responsible for passing laws. The ideas for these laws can come from a variety of places:

- A member of Congress
- Constituents
- Citizen's groups
- A member of the president's cabinet
- The president
- The executive agencies

The idea is then drafted into a bill. A bill should contain three things:

1. Statutory provisions, which describe what legislation will prohibit, what is required, etc.;
2. Administrative provisions, which describe who will be responsible for and enforce this statute, usually a department or agency; and
3. Judicial provisions, which describe which courts will handle disputes and aid enforcement.

These three provisions divide the responsibilities involved with the statute between the three branches.

## How a Bill Becomes a Law

There are six main steps a bill goes through on its way to becoming a law.

1. Introduction
2. Consideration in committee
3. Reintroduction
4. Debate in Congress
5. Presidential action
6. Enrollment

I. Introduction. This process can take place in either the House or the Senate.
   1. First reading. In the Senate, the bill is usually introduced by presenting it to the clerk at the presiding officer's desk. It can be introduced from the floor by a senator with a brief statement. In the House of Representatives the bill is simply dropped in a hopper and printed in the *Congressional Record.*
   2. The bill is numbered. Bills originating from the Senate are numbered S. (number), whereas bills from the House of Representatives are numbered H.R. (number).
   3. The bill is assigned to a standing committee for consideration.
   4. The bill is printed by the Government Printing Office.

II. Considered in Committee. Once in the standing committee, bills are often sent to a subcommittee. If it is a controversial or important topic, the subcommittee may hold a public hearing to get more information.

The subcommittee will then prepare a report for the standing committee with their recommendations and amendments.

The standing committee will then consider the subcommittee's report and take one of the several actions. It may

- pigeonhole the bill (kill the bill),
- report the bill out of committee favorably (recommend to the house of origination that it be passed),
- mark up the bill (amend the bill),
- throw out the old bill and write a new one, called a clean bill, or
- report the bill out unfavorably, which happens in rare cases when a committee has political reasons for not killing a bill.

Once the bill is reported out of the committee, it must then be placed on the calendar of the originating house. The Rules Committee decides exactly when and how the bill will be discussed.

III. Reintroduction (or calling up the bill). When it is the bill's turn on the calendar, it is reintroduced or called up by the standing committee to the full Senate or House. The bill is then considered reported out of committee.

IV. Debate in Congress. Once the bill is reintroduced, it has its second reading. At this time, the Congress members can take one of several actions on it. They can

- pass it as written
- table the bill. This removes the bill from further discussion, or kills it.
- send the bill back to committee. This often results in the bill being killed in committee.
- amend the bill. All amendments are debated and voted on.

All actions on the bill are published daily in the *Congressional Record.*

The bill then has its third reading and is voted on in its final form. If it gets a yes vote, it undergoes engrossment and graduates from being a bill to being an act. The act is then passed on to the other house, where it undergoes the same process, starting with the introduction of the act.

If the Senate and House of Representatives end up with different versions of the same act due to amendments, members are sent from each house to form a conference committee to resolve the differences.

A report of the committee's results is sent to each house, and the act is voted on again.

IV. Presidential action. The president has three options when presented with an act. The president can

A. approve and sign the act. It then becomes a law.
B. not act on it within 10 days, excluding Sundays. In that case it will automatically become a law unless Congress is out of session. In that case, it will be considered vetoed. This is called a *pocket veto*.
C. veto the act by not signing it or sending it back to the house of origin for recommendations. The veto can be overridden if each house approves it by a two-thirds majority.

Once the act becomes a law, it is printed as a slip law and is distributed to the public. If the law is a public law, then it is numbered consecutively with the Congress session number and the number of the law; for example, Pub. L. 104–4 is the fourth law passed by the 104th Congress.

V. Enrollment. The act is enrolled, or reprinted and submitted to the president after being signed by the Speaker of the House and the president of the Senate.

During every step of the legislative process, information about the bill or act is printed, as listed next.

1. Introduction: bills of resolutions
2. Considered in committee: hearings and reports
3. Calling up the bill: *Congressional Record*
4. Debate: *Congressional Record and Conference Report*
5. Presidential action: slip law or veto message
6. Enrollment: *U.S. Code*

Once the act becomes a law, it is then up to the regulatory agency to enforce it through the use of rules, regulations, policies, and guidelines.

In the past few years, it has become possible to find much of the legal information you need on the Internet. This is very convenient and allows you to get up-to-date information quicker. Searching these resources can sometimes be difficult, however, if you are unsure about what you are looking for. In these cases, it may be more helpful for you to use the conventional method. This involves using indexes and resources available from the federal government. Both the electronic means of finding legal information as well as the method of using indexes in the library will be discussed.

## Where to Find Legal Information

The federal government makes legal information available at federal depository libraries and on the Internet. There are approximately 1,400 federal depository libraries throughout the United States and its territories, at least one in every congressional district. All provide free public access to a wide variety of federal government information in both print and electronic formats and have expert staff available to assist users.

All of the information available on the Internet can be accessed from the U.S. Government Printing Office's GPO Access site (http://www.gpoaccess.gov/index.html). The Food and Drug Administration (FDA) has a helpful article online, How to Obtain FDA Statutes and Regulations (http://www.fda.gov/ora/fed_state/Small_business/sb_guide/howto.html).

Information about U.S. laws is printed in many different publications. It is important to know what each law contains and how often it is updated when looking for current information. Some governmental publications of interest are described below.

## Law-Making Process

Consult the following resources when looking for information about the actual passing of a law, for example, what took place during debate about the law or when the law will be introduced to the House of Representatives or the Senate.

*Congressional Record*. The *Congressional Record* is the official record of the proceedings and debates of the United States Congress. It is published daily when Congress is in session. Its index has two portions; the Index to the Proceedings and the History of Bills and Resolutions. It is published every two weeks.

*House Calendar*. The *House Calendar* contains a numerical order of bills and resolutions. In addition, a complete legislative history is given of each bill. It can be found on the Internet at the GPO Access site.

## The Laws

Once the laws have been passed, they can be found in the following forms.

### Slip Laws

Public laws are first printed as a slip laws. They can be found loose at federal depository libraries until the end of the year, when they are reprinted together as the *Statutes at Large*. They are referred to by their public law number (Pub. L. 104–4). GPO Access has the full text of public laws passed since the last Congress.

### The Statutes at Large

The *Statutes at Large* (Stat.) is the official compilation of federal laws. It is published annually. It contains all laws, both public and private, passed in the United States. At the end of each congressional session, acts are printed into the *Statutes at Large* in the order in which they are printed in the statutes.

At the beginning of each volume there is a list of bills enacted into public laws, a list of public laws by number, a list of proclamations, a popular names index, and a subject index. Because the text of laws published as public laws and *Statutes at Large* are the same, there is not a *Statutes at Large* database on GPO Access. However, users may perform a search by *Statutes at Large* citation in both the public laws and *U.S. Code* databases. The *Statutes at Large* are can be found in federal deposit libraries.

### U.S. Code

The *U.S. Code* is the codification by subject matter of the general and permanent laws of the United States based on what is printed in the *Statutes at Large*. It is divided by broad subjects into 50 titles and published by the Office of the Law Revision Counsel of the U.S. House of Representatives. Of the 50 titles, only 23 have been enacted into positive (statutory) law. These titles are 1, 3, 4, 5, 9, 10, 11, 13, 14, 17, 18, 23, 28, 31, 32, 35, 36, 37, 38, 39, 44, 46, and 49. When a title of the code was enacted into positive law, the text of the title became legal evidence of the law. Titles that have not been enacted into positive law are only *prima facie* evidence of the law. In that case, the *Statutes at Large* still govern.

The *U.S. Code* also contains helpful indexes and tables. The General Index contains an alphabetical listing of useful subject headings. Entries are also listed under agency names. The *U.S. Code* citation is given for each entry.

The *U.S. Code* has nine tables found at the end of its volumes. Two tables of particular interest are discussed next.

### Table 1—Revised Titles

This table lists all the sections of the titles that have been revised since the last printing.

### Table 3—Statutes at Large

Here all the public laws currently in effect and their corresponding *U.S. Code* citation are listed.

Statutes are cited in the *U.S. Code* as Title *U.S.C.* section (subsection), for example*: 21 U.S.C. Sec. 301 (a).*

A *U.S. Code* supplement is issued during each of the years between printings of the *U.S. Code*. This contains additions to and changes in the general and permanent laws of the United States enacted during that Congress and session.

The *U.S. Code* does not include regulations issued by executive branch agencies, decisions of the federal courts, treaties, or laws enacted by state or local governments. Regulations issued by executive branch agencies are available in the *Code of Federal Regulations* (*CFR*). Proposed and recently adopted regulations may be found in the *Federal Register*.

Since 1926, the *U.S. Code* has been published every six years. Between editions, annual cumulative supplements are published to present the most current information.

When a section is affected by a law passed after a supplement's revision date, the header for that section includes a note that identifies the public law affecting it. To find the updated information, you must search the public laws databases for the referenced public law number.

The *U.S. Code* can be searched on the Internet from GPO Access (http://www.gpoaccess.gov/uscode/index.html). You can search the *U.S. Code* by subject or by citation. GPO Access contains the 2000 and 1994 editions of the *U.S. Code*, plus annual supplements. The *U.S. Code* on GPO Access is the official version of the code; however, two unofficial editions are available. These are the *U.S.C.A.*(*U.S. Code Annotated*) and the *U.S.C.S.* (*U.S. Code Service*). The *U.S.C.A.* and *U.S.C.S.* contain everything that is printed in the official *U.S. Code* but also include annotations to case law relevant to the particular statute. While these unofficial versions may be more current, they are not official and not available from the U.S. Government Printing Office.

### Code of Federal Regulations

The *Code of Federal Regulations* (*CFR*) is the codification of the general and permanent rules published in the *Federal Register* by the executive departments and agencies of the federal government. It is divided into 50 titles that represent broad areas subject to federal regulation. Each title is divided into chapters, which usually bear the name of the issuing agency. Each chapter is further subdivided into parts that cover specific regulatory areas. Large parts may be subdivided into subparts. All parts are organized in sections, and most citations in the *CFR* are provided at the section level. Titles 7 (Agriculture) and 21 (Food and Drug) contain most laws concerning food.

Each volume of the *CFR* is updated once each calendar year and is issued on a quarterly basis.

- Titles 1–16 are updated as of January 1st.
- Titles 17–27 are updated as of April 1st.
- Titles 28–41 are updated as of July 1st.
- Titles 42–50 are updated as of October 1st.

*CFR* volumes are added to GPO Access concurrent with the release of the paper editions. When revised *CFR* volumes are added, the prior editions remain on GPO Access as a historical set. Some *CFR* records on GPO Access date back to 1996; all titles are available from 1997 to the current year.

Due to the update schedule of the *CFR*, the List of Sections Affected (LSA) provides a cumulative list of *CFR* sections that have been changed at any time since each *CFR* title was last updated.

## Conventional Search Method

To find a specific law using printed material, use the following methods.

1. If the name of the law is known, look in the *U.S. Code* Popular Names Index. This index contains public law citations and *U.S. Statutes at Large* citations for many laws. The Popular Names Index is found in

the same volume as Title 50 in the 1994 edition. If the name of the law is not known, look in the *U.S. Code* General Index for laws about a specific subject.

2. Find out the status of the law by using the statutes at large table in the *U.S. Code* tables volume. This table lists laws by their public law number and gives their *U.S. Code* citation and their status.
3. Take the *U.S. Code* citation given and use it to find the law.

## Example: Locate the Nutrition Labeling Law

### Conventional Paper Search

1. Look in the *U.S. Code* under the General Index for a nutrition labeling.
2. There you find the entries for Nutrition Labeling and Education Act Amendments of 1993 and the Nutrition Labeling and Education Act of 1990. Under both of these entries it says: "Text of Act. See Popular Name Table."
3. Go to the Popular Name Table and look under Nutrition Labeling and Education Act of 1990. This lists four public law citations.
4. To find their *U.S. Code* citation, look in the *U.S. Code Statutes at Large* table. There you will find the *U.S. Code* citation for each of the public laws.
5. To find this law, look in Title 21 Section 301nt of the *U.S. Code*. This turns out to be the Federal Food, Drug, and Cosmetic Act (FFDCA).
6. Section 301 is a list of short titles of acts that amend the FFDCA, and Pub. L.101–535 is listed there as the Nutrition Labeling and Education Act. It also lists all parts of the FFDCA that are amended by the NLEA.

This list or the one found in the U.S. Statutes table can be used to find all parts of the law in Title 21 of the *U.S. Code*.

### Using the Internet

1. First go to United States Code (http://www.gpoaccess.gov/uscode/index.html). Type in "nutrition labeling" at the Search box, using the quotation marks, and click on Submit.
2. The third hit says "21 USC Sec. 301. Short Title". Click on TEXT to read it.
3. Sec. 301 says "Short Title: This chapter may be cited as the Federal Food, Drug, and Cosmetic Act." The Nutrition Labeling and Education Act is an amendment to the Federal Food, Drug and Cosmetic Act. . .

4. Go down to the section titled "Short Title of 1990 Amendments." There you will see that the last paragraph refers to the Nutrition Labeling and Education Act and that each section that is amended by the act is listed. You can then search for these sections to read them. For example, to read section 341–1, type in "21USC341".

The GPO Access version of the *U.S. Code* offers a browse function. Browse is one of the database functions that can be found in the left column of the *U.S. Code* main page (http://www.gpoaccess.gov/uscode/index.html). The browse function of the *U.S. Code* provides the list of titles, and by clicking on Title you will get a list of chapters within a title.

## How Regulations Are Made

### Making Rules and Regulations

After the bill becomes a law, the agency or department listed in the administrative provisions is responsible for enforcing the statute. To do this, the agency must make rules and regulations. Rules are typically administrative in nature, while regulations deal more with scientific and technical issues. They both carry the force of the law, which means that if a rule or regulation is broken, then the statute is broken. For the purpose of this discussion, the terms *rule* and *regulation* will be used interchangeably.

Regulations are made in the following manner.

1. When a regulation is proposed by the appropriate agency it is then published in the *Federal Register*, which is the federal publication that notifies the public of changes in U.S. laws and regulations. It is published as an Advanced Notice of Proposed Rulemaking, or ANPR. ANPRs alert the public that the agency plans to make or change a regulation and asks for their comments.
2. There is a comment period during which people can write in with opinions and criticisms of the proposed regulation. These comments are taken into account when the final regulation is made, and a summary of the comments is provided in the final rule. In addition, the agencies can have hearings where they bring in experts on a particular subject to aid their decision.
3. The regulation is printed in the *Federal Register* as a final rule along with the date it goes into effect.

A list of where each step in this process is published can be found next.

The Rule-Making Process and Publication

1. A rule is proposed in the Proposed Rules section of the *Federal Register.*
2. A comment period is announced in the Proposed Rules section of the *Federal Register.*
3. The final rule is published in the Final Rule section of the *Federal Register.*
4. Rules are arranged according to subject in the *CFR*. Regulations are compiled into the *CFR*.

### Rule-Making Example

An example of this process is the Pathogen Reduction; Hazard Analysis Critical Control Point Systems Final Rule that the U.S. Department of Agriculture (USDA). The public was first notified that the USDA was planning a new inspection system when the Advanced Notice for Proposed Rulemaking (ANPR) was published in the *Federal Register* of December 29, 1995 (60FR 67469). This ANPR explained the regulation and requested comments from the public. There was an initial comment period of 120 days. The comment period was ultimately extended for 30 days, then reopened for another 95 days.

During this time, seven informational briefings were held in seven cities around the United States to help interested parties prepare comments on the proposal. A panel of FSIS officials and scientists provided information on the proposed regulations and answered questions. All of this input assisted the USDA in writing the final rule.

Anyone can comment on a proposal, and comments do make a difference. To learn more about this process, read Making Your Voice Heard at FDA: How to Comment on Proposed Regulations and Submit Petitions at http://www.fda.gov/opacom/backgrounders/voice.html.

## Code of Federal Regulations

All the final and interim regulations of the United States are compiled (or *codified*) into the *CFR*. The *CFR* is updated yearly. The regulations are categorized into 50 titles. (Note: These are not the same titles as those in the *U.S. Code*—that would be too easy!) These titles are subdivided into chapters, parts, sections, and paragraphs. Regulations are referred to using those parts. An example citation for a regulation in the *CFR* is 21 *CFR* Sec. 131.144 (a). This would be read as Title 21, part 131, section 144, paragraph a.

Titles of particular interest to food scientists are Title 9, Animals and Animal Products, and Title 21, Food & Drugs.

The *CFR* can be found on the Internet at the GPO Access site. The best way to become familiar with how these regulations are printed is to jump right in and look at them.

Published along with the *CFR* are the *CFR* Index and Finding Aids. These resources are available to help people find information located in the *CFR*. The *CFR* Index has appropriate *CFR* citations under subject headings and agency names.

## Parallel Table of Authorities and Rules

This table lists rule-making authority for regulations codified in the *CFR*. It contains a section for *U.S. Code* citations, *U.S. Statutes at Large* citations, public law citations, and presidential document citations. Within each segment the citations are arranged in numerical order.

Also, the list of *CFR* titles, chapters, subchapters, and parts outlines what can be found in each section of the *CFR*.

## Alphabetical List of Agencies Appearing in the CFR

Each agency is listed along with the *CFR* title, subtitle, or chapter in which it is found.

The *CFR* also publishes a monthly publication, *The List of CFR Sections Affected* (*LSA*), which lists the sections of the *CFR* that have been changed by actions published in the *Federal Register*. This includes any new or proposed rules. Entries are by *CFR* title, chapter, part, and section. The *LSA* should be consulted whenever looking for up-to-date regulations to ensure the *CFR* has not been changed. The *Federal Register* contains a cumulative list of parts affected. The page numbers to the right indicate where the specific amendments begin in the *Federal Register*. A table of *Federal Register* issue pages and dates can be found at the back of the *LSA*.

### Using the CFR

To find a regulation in the *CFR*, do the following.

1. Look up the subject of the regulation in the *CFR* Index and Finding Aids.
2. Go to the source listed.
3. Check the *LSA* or the *Federal Register* for any recent changes to the regulation.

### Using the Internet

The *CFR* can be searched on the Internet from the GPO Access, Code of Federal Regulations. Here you can search selected books and titles of the *CFR*, search by keywords, or search by citation.

After locating the *CFR* reference, you should check the federal regulations to see if there have been any changes.

1. Go to the *Code of Federal Regulations.*
2. If you know the title of the *CFR* that the regulation should fall under, you can specify it by clicking on Search Your Choice of *CFR* Books, Available Online. (Usually food regulations will be under Title 9. Animal/Animal Products or 21. Food and Drugs.)
   Click on Continue.
3. You will then get a table with the titles you selected with all the volumes in that title. If you know the volume where your regulation is published, you can narrow down your search by clicking on that volume.
4. At the Search Terms box enter your subject and hit Submit.
5. You will then get a listing of relevant regulations. You can look at a summary, the text, or a pdf graphic of the law.
6. Click on Search the Federal Register for Related Documents to see if there have been any changes in the regulations.

## References

USIA. *The Legislative Branch: The Reach of Congress.*

Friedman, Lawrence M. 1984. *American Law.* New York: W.W. Norton and Company.

Ferguson, John H. and Dean E. McHenry. 1969. *The American System of Government.* New York: McGraw-Hill Book Company.

Glossary of Terms. Understanding the Federal Courts. http://usinfo.org/law/courts/fr899_toc.htm (June 2004).

The Constitution of the United States of America. United States Information Agency. Online. http://usinfo.state.gov/topical/rights/structur/constitu.htm (June 2004).

Hardy, Richard J. 1994. *Government in America.* Boston: Houghton Mifflin Company.

We the People: The Structure of the U.S. Government. United States Information Agency. Online. http://usinfo.state.gov/topical/rights/structur/main2.htm (June 2004).

An Outline of American Government. U.S. Information Agency. Online. http://usinfo.state.gov/usa/infousa/politics/govworks/oagtoc.htm.

The Executive Branch: Powers of the Presidency. An Outline of American Government. United States Information Agency. http://usinfo.state.gov/usa/infousa/politics/govworks/oag-pt3.htm (June 2004).

The Federal Judicial Branch. Understanding the Federal Courts. http://usinfo.org/law/courts/fr899_toc.htm (June 2004).

The Judicial Branch: Interpreting the Constitution. An Outline of American Government. The United State Information Agency. Online. http://usinfo.state.gov/usa/infousa/politics/govworks/oag-pt5.htm (June 2004).

The Legislative Branch: The Reach of Congress. An Outline of American Government. The United States Information Agency.

## Additional Resources

The following list provides resources for more information on the U.S. government.

Interactive Citizen's Handbook
http://www.whitehouse.gov/government/handbook/index.html

Basic Readings in U.S. Democracy
http://usinfo.state.gov/usa/infousa/facts/democrac/demo.htm

The Judicial Branch: Interpreting the Constitution. U.S. Information Agency. Online.
http://usinfo.state.gov/usa/infousa/politics/govworks/oag-pt5.htm

U.S. House of Representatives home page
http://www.house.gov/

The United States Senate World Wide Web Server
http://www.senate.gov/

The U.S. Courts home page. http://www.uscourts.gov/

The White House
http://www.whitehouse.gov/

The Legislative Process. U.S. House of Representatives
http://www.house.gov/house/Tying_it_all.html

Federal Depository Library Locator. GPO Access
http://www.gpoaccess.gov/libraries.html

How to Obtain FDA Statutes and Regulations. Office of Regulatory Affairs
http://www.fda.gov/ora/fed_state/Small_Business/sb_guide/howto.html
GPO Accesshttp://www.gpoaccess.gov/multidb.html

This is the best starting place when looking for government information. Over 70 databases can be accessed from this site, including the *Federal Register, U.S. Code*, and the *Congressional Record.*

Keeping America Informed: United States Government Printing Office.
http://www.gpoaccess.gov/index.html

This site links to government information, including GPO Access. The U.S. Government Printing Office disseminates official information from all three branches of the federal government.

Specialized Search Pages for Databases Online via GPO Access
http://www.gpoaccess.gov/databases.html

This list links to specialized searches for many databases, including the *CFR* and the *Federal Register*.
*Code of Federal Regulations (CFR)* http://www.gpoaccess.gov/cfr/index.html

This site hosted by the National Archives and Records Administration allows for specialized searches of the *CFR*.

Congressional Record Online via GPO Access
http://www.gpoaccess.gov/crecord/index.html

This site allows you to search the 1995, 1996, and 1997 *Congressional Record* (Volumes 141, 142, and 143) by section.

*Federal Register* Online via GPO Access
http://www.gpoaccess.gov/fr/index.html

Database for the 1995, 1996, and 1997 *Federal Register* (Volumes 60, 61 and 62). Also includes a link to helpful hints for searching the *Federal Register*.

The United States Code http://www.law.cornell.edu/uscode/index.html

Cornell University has the *U.S. Code* available online in a easy-to-read format.

History of Bills Online via GPO Access
http://www.gpoaccess.gov/hob/index.html

The History of Bills and Resolutions is a section of the *Congressional Record* Index that provides information about all bills and resolutions introduced during that session of Congress. This database allows searches of the History of Bills from 1983 to 1997.

CHAPTER 2

# How Did We Get Where We Are Today?

Patricia Curtis, Auburn University
Wendy Dunlap

Throughout history there have been food laws. Ancient Samarians had a statute that required innkeepers to give their customers the proper amount of beer or "her" hand would be cut off.

The Bible contains many references to food and food laws. The majority of these laws were passed down from Moses and are the basis for Kosher foods in the Jewish faith. These laws dealt with clean versus unclean animals and prohibited practices. They also address the use of just balances and weights in the marketplace.

The adulteration of food was also addressed in many treatises in ancient times. For example, Theophrastus (370–285 BC) wrote the botanical treatise *Enquiry into Plants*. This work discussed the use of artificial preservatives and flavors, such as balsam gum, that were added to many foods for economic reasons.

*Caveat emptor* or "the buyer beware" was the idea behind Roman civil law. The penalty for selling adulterated food in Rome in AD 400 was banishment from Rome or slavery.

In the 1600s, London had laws in place against food adulterations, and local guilds enforced their own rules regarding unfair practices. In more recent European history, there were many treatises written about food adulteration. One German treatise written in 1820 taught housewives how to test their food for adulteration and became a best seller.

## Reasons for Food Laws

Food laws were made for a variety of reasons, including

- to ensure that consumers get what they pay for,
- to ensure that the food is safe, and
- to uphold religious beliefs.

The first two reasons listed above deal with the concepts of *adulteration* and *misbranding*. Adulteration and misbranding are strictly defined in American food laws. According to *The Random House Dictionary of the English Language*, to adulterate is defined as follows: to debase by adding inferior materials or elements; make impure by admixture; use cheaper, inferior, or less desirable goods in the production or marketing of any professedly genuine article.

Adulteration is the act or process of adulterating. When the adulterant does not actually physically harm someone, but cheats them out of their money, this is called *economic adulteration*. Examples include adding water to wine or milk, ash to pepper, and chalk to bread. Adding dyes to conceal inferior ingredients is also considered adulteration.

To misbrand something is to "brand or label erroneously." All of the American food laws are based on prohibiting these two activities.

## American Food Laws

When the settlers came to the United States they brought with them European ideas of food laws and regulations. Although most food was sold in bulk, there were a few processed foods available, most importantly, bread. Bread has been regulated throughout the ages, mostly to assure that it was affordable and of a suitable quality.

The first food laws and regulations in the United States were passed by local communities or states and were loosely based on regulations in England. The Massachusetts Bay Colony passed the first food adulteration law in 1641, which protected meat and pork, and the first colonial Assize of Bread was enacted in 1646. This assize was almost identical to the English Assize and required bakers to label their bread and set required weights for three qualities of bread. In addition, each town was required to have two bread inspectors to ensure that the bakers were obeying the law.

As the United States became increasingly more urban and industrialized, fewer people grew their own food. Therefore, marketplaces were developed where people could purchase the food they needed. As more food was produced to meet demand, adulteration became a greater problem. Increases in technological knowledge led to many new ingredients and products, many of which were unsafe. Thus, not only was economic adulteration a problem, but adulteration leading to unsafe products increased as well. While most states had regulations in place to deal with adulteration, they did not apply to products involved in interstate commerce.

In 1880 the chief chemist of the USDA, Peter Collier, recommended the enactment of a national food and drug law. Congress was reluctant to comply, however, because many Americans felt that food regulation was a state issue and that the federal government did not have the right to legislate this type of law.

Meanwhile, the New York State Board of Health reported data (Tables 2.1 and 2.2) from studies conducted in 1882 to determine the level of adulteration occurring in food in New York.

At that time, the safety of foods was the responsibility of the Bureau of Chemistry, located in the U.S. Department of Agriculture (USDA). The Bureau of Chemistry is the predecessor to the Food and Drug Administration (FDA).

**Table 2.1.** Level of adulteration in New York in 1882

| *Article* | *Number of Samples Tested* | *Number Found To Be Adulterated* | *Percent Adulterated* |
|---|---|---|---|
| Butter | 40 | 21 | 52.5 |
| Olive oil | 16 | 9 | 56.3 |
| Baking powder | 84 | 8 | 9.5 |
| Flour | 117 | 8 | 6.8 |
| Spices | 180 | 112 | 62.2 |
| Coffee (ground) | 21 | 19 | 90.5 |
| Candy (yellow) | 10 | 7 | 70.0 |
| Brandy | 25 | 16 | 64.0 |
| Sugar | 67 | 4 | 6.0 |

*Source:* Taken from Battershall, Jesse P. *Food Adulteration and Its Detection.* New York: E. & F. N. Spon, 1887.

**Table 2.2.** Common food adulterants in 1887

| *Food Article* | *Common Adulterants* |
|---|---|
| Baker's chemicals | Starch, alum |
| Bread and flour | Other meals, alum |
| Butter | Water, coloring matter, oleomargarine, and other fats |
| Canned foods | Metallic poisons |
| Cheese | Lard, oleomargarine, cottonseed oil, metallic salts (in rind) |
| Cocoa and chocolate | Sugar, starch, flour |
| Coffee | Chicory, peas, rye, corn, coloring matters |
| Confectionary | Starch-sugar, starch, artificial essences, poisonous pigments, terra alba, plaster of Paris |
| Honey | Glucose syrup, cane sugar |
| Malt liquors | Artificial glucose and bitters, sodium bicarbonate, salt |
| Milk | Water and removal of cream. |
| Mustard | Flour, turmeric, cayenne |
| Olive oil | Cottonseed and other oils |
| Pepper | Various ground meals |
| Pickles | Salts of copper |
| Spices | Pepper dust, starch, flour |
| Spirits | Water, fusil oil, aromatic ethers, burnt sugar |
| Sugar | Starch-sugar |
| Tea | Exhausted tea leaves, foreign leaves, indigo, Prussian blue, gypsum, soapstone, sand |
| Vinegar | Water, sulfuric acid |
| Wine | Water, spirits, coal tar and vegetable colors, factitious imitations |

*Source:* Taken from Battershall, Jesse P. *Food Adulteration and Its Detection.* New York: E.& F.N. Spon, 1887.

## The "Poison Squad"

In the past, most adulteration was classified as economic adulteration. However, toward the turn of the century, the use of potentially harmful adulterants in food increased considerably. The use of harmful food additives came under considerable scrutiny in 1902 when the chief chemist of the Bureau of Chemistry, Harvey W. Wiley, began studying their effects on humans. Twelve young men volunteered to be "guinea pigs" in testing the effects of additives that Dr. Wiley believed to be dangerous. These additives included formaldehyde, boric acid and borax, salicylic acid and salicylates, sulfurous acid and sulfites, and benzoic acid and benzoates. The men were fed a natural diet plus one of the additives of interest.

Once Americans were informed of the "Poison Squad," named by a newspaper reporter, they could not get enough information about how the experiment was proceeding. Many stories about Dr. Wiley and his volunteers were published in the newspapers and magazines of the time and thus brought the food additives issue to the attention of American consumers.

In response to the heightened public awareness and the findings of the Poison Squad study, President Theodore Roosevelt recommended on December 5, 1905, that a law be enacted to regulate interstate commerce of adulterated and misbranded foods, drinks, and drugs. There was a tremendous response by consumers to their senators and representatives, which led to the passage of the Pure Food and Drug Act on June 30, 1906.

## The Jungle

Earlier that same year, Upton Sinclair published *The Jungle*. While this book was meant to be a motivation for adopting socialism, it actually became regarded as a criticism of the meat industry and fueled the efforts for the enactment of the 1906 Pure Food and Drug Act. In the book Sinclair describes the conditions in Chicago's meat-processing plants.

> There was never the least attention paid to what was cut up for sausage; there would come all the way back from Europe old sausage that had been rejected, and that was moldy and white—it would be dosed with borax and glycerine, and dumped into the hoppers, and made over again for home consumption. There would be meat that had tumbled out on the floor, in the dirt and sawdust, where workers had tramped and spit uncounted billions of consumption germs. There would be meat stored in great piles in rooms; and the water from leaky roofs would drip over it, and thousands of rats would race about it. It was too dark in these storage places to see well, but a man could run his hand over these piles and sweep off handfuls of the dung of rats. These rats were nuisances, and the

packers would put poison bread out for them, they would die, and then rats, bread, and meat would go into the hoppers together.

In direct response to this book, the meat-packing houses in Chicago were eventually inspected, and it was determined that the conditions described in the book were accurate. At President Roosevelt's urging, the Federal Meat Inspection Act was passed in 1906 on the same day as the Food and Drug Act. This act was to be administered by the Bureau of Animal Industry housed in the USDA.

## The Need for a New Food and Drug Law

Although the 1906 Pure Food and Drug Act was a step in the right direction, it did have weaknesses. These weaknesses were outlined in a 1917 report by the USDA Bureau of Chemistry and included:

- the lack of legal standards (descriptions) for foods,
- the lack of authority to inspect food and drug warehouses,
- the inability to restrict the interstate shipment of a food that naturally contain poison, and
- lack of jurisdiction over false or misleading claims made on food.

Another weakness of the 1906 law was its failure to address pesticide residues in foods. This weakness became evident in the early 1920s when Britain threatened to stop importing American apples because they contained a level of lead arsenic, a pesticide that caused illness. Growers responded by washing apples intended for export in order to remove the residue. Apples intended for sale in the United States, however, were not washed.

When the Bureau of Chemistry reacted by establishing a tolerance for lead arsenic on apples sold in the United States, the growers strongly protested. Throughout a 17-year period, the USDA tried to enforce lower lead arsenic tolerances. It was unable to do so because the 1906 law did not grant them that authority. The only way it could justify the tolerance was through the expensive and time-consuming process of presenting scientific evidence of lead arsenic's toxicity in court proceedings when sued.

## Elixir Sulfanilamide

There were also weaknesses in the 1906 law that involved drugs. One weakness was the lack of mandatory premarket drug testing. This failure of the law was tragically exposed in the elixir sulfanilamide incident. In the early 1930's, a manufacturer prepared a liquid form of sulfanilamide using diethylene glycol. While sulfanilamide itself is not harmful, diethylene glycol is toxic. As a result of the new drug, over 100 people died.

In addition to the well-publicized events described above, the public became increasingly aware of the problems still present in the regulation of food and drugs through the publication of several informative if provocative books in the 1930s; *100,000,000 Guinea Pigs* by A. Kallet and F.J. Schlink, *American Chamber of Horrors: The Truth about Food and Drugs* by R. Lamb, and *Guinea Pigs No More* by J.B. Matthews are just a few examples of these books.

## Food, Drug, and Cosmetic Act of 1938

The increasing public concern over food safety, the inability of the USDA to adequately regulate food and drugs under the 1906 law and the horror of the sulfanimide tragedy acted as catalysts for motivating Congress to pass the Federal Food, Drug, and Cosmetic Act (FDCA) of 1938. New provisions of the act included

- providing that safe tolerances be set for unavoidable poisonous substances,
- authorizing standards of identity, quality, and fill-of-container for foods,
- authorizing factory inspections,
- adding the remedy of court injunctions to the previous penalties of seizures and prosecutions, and
- expanding the definitions of adulteration and misbranding.

## Food Additives

Even after the passage of the 1938 FDCA, concern for food additives continued among scientists and the public. In 1951, a House committee under the direction of Chairman James Delaney began examining the safety of food additives and pesticides. As a result, the Pesticide Amendment of 1954, the Food Additive Amendment of 1958, and the Color Additive Amendment of 1960 were passed by Congress. Each of these amended the 1938 FDCA.

## *Silent Spring* and the Environmental Protection Agency

In 1962, Rachel Carson's book, *Silent Spring,* was published. This book was the first to examine the effects of widespread use of pesticides and other chemicals on the environment. In her book she described a town where all life had been killed due to the overuse of chemicals. The book quickly polarized the nation, dividing people between those who accepted the author's concerns and were her advocates and those who rejected the book as fictitious. Whether it was endorsed or rejected, the book did have an impact.

*Silent Spring* also described how pesticide regulation was influenced by the agricultural chemical industry. This publicity, along with pressure from the new environmental activists, led to the formation of a new national agency, the Environmental Protection Agency (EPA), which would report directly to the president and would be responsible for all issues dealing with the environment. One of its responsibilities was to relieve the FDA from establishing pesticide tolerances for food. The EPA was formed in 1970.

## A Brief History of Agriculture

Many government agencies begin as part of USDA: the FDA, the EPA, the Biological Survey of the Department of the Interior, the Forest Service, and the Soil Conservation Service.

In 1776 George Washington suggested to Congress the formation of the National Board of Agriculture. It was 1819 before New York formed the first State Board of Agriculture, however. By 1820 agriculture began to demand a place in government and the U.S. House of Representatives formed the first Agriculture Committee. It was not until 1825 that the Senate formed its first Agriculture Committee. In 1839, the Agricultural Department of the U.S. Patent Office was established to collect statistics, distribute seeds and plants and compile and distribute information.

### U.S. Department of Agriculture

In 1862, the Agricultural Department of the U.S. Patent Office was established as an independent entity, the USDA. The Agriculture Building was constructed in 1867 on the Mall in Washington, DC. This building was torn down in the 1930s to make room for the current USDA complex on Independence Avenue. The Bureau of Home Economics was established in 1923 under the Division of Food and Nutrition. Various surveys were conducted to compare rural and urban living. In 1946, the National School Lunch Act was passed. The National School Lunch Program established standards that provided a well-balanced meal to meet one-third of a child's daily dietary allowance.

In 1862, when President Abraham Lincoln founded the USDA, he called it the "people's department." In Lincoln's day, 48 percent of the population were farmers who needed good seeds and information to grow their crops. By 1996 farm families made up less than 10 percent of rural populations. However, new records were set in 1996 when the net farm income exceeded $51 billion and agricultural exports reached $59.8 billion.

### State Departments of Agriculture

New York appointed its first state entomologist in 1853. A few states began to inspect dairy products in the 1870s; however, it was not until 1874 that Georgia set up the first state department of agriculture.

### The Environmental Protection Agency

The U.S. EPA was established in 1970 to consolidate into one agency a variety of federal research, monitoring, standard-setting, and enforcement activities to ensure environmental protection. The EPA's mission is to protect human health and to safeguard the natural environment—air, water, and land.

### Brief History of the FDA

The agency that is now the FDA started out as the Division of Chemistry in the Department of Agriculture and has been under a number of different federal departments and titles since its origin. Table 2.3 traces its development up to the present.

## Current Consumer and Regulatory Concerns

The safety of food additives has remained a concern throughout the decades. This is evident by the many books that have been published about the subject, most of them written for the average consumer.

**Table 2.3.** Development of the FDA

| | |
|---|---|
| 1862 | The Division of Chemistry is formed under the Department of Agriculture (USDA). |
| 1901 | The Division of Chemistry is renamed as the Bureau of Chemistry. |
| 1927 | The Bureau of Chemistry is renamed as the Food, Drug and Insecticide Administration. |
| 1930 | The Food, Drug and Insecticide Administration is renamed as the Food and Drug Administration (FDA). |
| 1940 | The FDA is moved from the USDA to the Federal Security Agency. |
| 1953 | The Federal Security Agency becomes the Department of Health, Education and Welfare (HEW). |
| 1979 | The FDA is moved to the Department of Health and Human Services. |
| 1988 | The Food and Drug Administration Act established the FDA as an agency under the Department of Health and Human Services. |

Recently, due to many well-publicized outbreaks, pathogenic food microorganisms have became an increasing concern in the public's eye. Outbreaks from *Escherichia coli* O157:H7 in hamburger and fresh produce, bovine spongiform encephalitis (mad cow disease) in beef, *Listeria* in ready-to-eat foods such as cheese and sandwich meat, and *Salmonella* on poultry and raw eggs are all recent concerns that are receiving a lot of attention. Regulators are responding with new microbial standards and mandatory testing. Meat and poultry inspection is being shifted toward a more science-based method through the use of Hazard Analysis Critical Control Point (HACCP) systems, and FDA-regulated foods are moving in that direction, as well.

# Major Food Laws

## Pure Food and Drug Act of 1906

The Food and Drug Act of 1906 prohibited the movement of misbranded and adulterated foods, drinks, and drugs in interstate commerce. It allowed the seizure of adulterated or misbranded products and allowed criminal penalties if the law was broken. It did not, however, provide any food standards, address false advertising of food or drugs, allow inspection of food or drug warehouses, or address foods in which a poisonous substance occurs naturally.

## Federal Meat Inspection Act of 1906

The Meat Inspection Act of 1906 gave the USDA authority to inspect any meat intended for interstate commerce or export. It forbids the adulteration or misbranding of meat products.

## The Food, Drug, and Cosmetic Act of 1938

This is the most important law in food and drug regulation today. When passed in 1938, it addressed the weaknesses in the 1906 law and has had many amendments since then. New provisions in the FDCA that regulate food include the following actions:

- providing that safe tolerances be set for unavoidable poisonous substances,
- authorizing standards of identity, quality, and fill-of-container for foods,
- authorizing factory inspections, and
- adding the remedy of court injunctions to the previous penalties of seizures and prosecutions.

## Poultry Products Inspection Act of 1957 (as amended 1968)

The Poultry Products Inspection Act regulates poultry products in a manner similar to that of the Meat Inspection Act.

## Food Additive Amendment of 1958

This amendment to the FDCA requires that new food additives be approved by the FDA and that their safety be proved by the manufacturer. One important provision of this amendment is called the Delaney clause. This provision prohibits the addition of any food that causes cancer.

## Color Additive Amendment of 1960

The Color Additive Amendment improved the FDCA. It is similar to the Food Additive Amendment in that it requires manufacturers to establish the safety of color additives in foods, drugs, and cosmetics.

## Fair Packaging and Labeling Act of 1966

The Fair Packaging and Labeling Act requires truthful and informative labeling of packages that are sold through interstate commerce.

## Egg Products Inspection Act of 1970

The Egg Products Inspection Act gave the USDA the responsibility of regulating egg products.

## Nutrition Labeling and Education Act of 1990

Amending the FDCA of 1938, the Nutrition Labeling and Education Act (NLEA) requires specific nutrition information to appear on food labels.

## Dietary Supplement Health and Education Act of 1994

This act defined *dietary supplements* and *dietary ingredients* and classified them as food. The act also established a commission to recommend how to regulate claims.

## Saccharin Notice Repeal Act of 1996

This act repealed the saccharin notice requirements.

## Food Quality Protection Act of 1996

Amended the FDCA, eliminating application of the Delaney proviso to pesticides.

## The Bioterrorism Act of 2002

The events of September 11, 2001, reinforced the need to enhance the security of the United States. Congress responded by passing the Public Health Security and Bioterrorism Preparedness and Response Act of 2002 (the Bioterrorism Act). The Bioterrorism Act is divided into five titles:

1. Introduction
2. Title I—National Preparedness for Bioterrorism and Other Public Health Emergencies
3. Title II—Enhancing Controls on Dangerous Biological Agents and Toxins
4. Title III—Protecting Safety and Security of Food and Drug Supply
5. Title IV—Drinking Water Security and Safety
6. Title V—Additional Provisions

The FDA is responsible for carrying out certain provisions of the Bioterrorism Act, particularly Title III, Subtitle A (Protection of Food Supply) and Subtitle B (Protection of Drug Supply).

A few key events in the history of food laws in this country were discussed here. Much more history is available, and many of the references listed give spirited descriptions of the politics and personalities involved.

# References

Battershall, Jesse P. 1887. *Food Adulteration and Its Detection.* New York: E. & F.N. Spon.

Bruce, Edwin, M. 1971. *Detection of the Common Food Adulterants.* New York: D. Van Nostrand Company.

Burditt, George M. 1995. The history of food law. *Food Drug Cosmetic Law Journal* Anniversary Issue.

Economic Research Service. 1997. A history of American agriculture 1776–1990. *United States Department of Agriculture.* Online http://www.usda.gov/history2/text9.htm (September 2004).

EPA. 1992. *The Guardian: Origins of the EPA*. EPA Historical Publication 1, Spring 1992. www.epa.gov/history/publications/print/origins.htm.

FDA. 1995. Milestones in U.S. Food and Drug law history. *FDA Backgrounder.* Online http://vm.cfsan.fda.gov/mileston.html (September 2004).

FDA. 1995. Selected sources on the history of FDA. *FDA Backgrounder.* Online http://www.fda.gov/opacom/morechoices/sources.html (September 2004).

Hutt, Peter Barton. 1982. Food and drug law: A strong and continuing tradition. *Food Drug Cosmetic Law Journal* 37:123–137.

Hutt, Peter Barton, and Richard A. Merrill. 1991. *Food Law and Drug Law: Cases and Materials.* Second Edition. Westbury, New York: The Foundation Press, Inc.

Hutt, Peter Barton, and Peter Barton Hutt, II. 1984. A history of government regulations of adulteration and misbranding of food. *Food Drug Cosmetic Law Journal.* 39:2–73.

Janssen, Wallace F. 1975. America's first food and drug laws. *Food Drug Cosmetic Law Journal.* 30:665.

Linton, Fred B. 1949. Federal food and drug laws—Leaders who achieved their enactment and enforcement. *Food Drug Cosmetic Law Quarterly.* 4:451.

Roberts, Howard R. 1981. Food safety in perspective. In *Food Safety.* New York: John Wiley and Sons. Chapter 1.

Stein, Jess, Ed. 1966. *The Random House Dictionary of the English Language: The Unabridged Edition.* New York: Random House.

Vetter, James L. 1966. *Food Laws and Regulations.* Manhattan, KS: American Institute of Baking.

CHAPTER 3

# Federal, State, and Local Laws

Patricia Curtis, Auburn University
Wendy Dunlap

Initially U.S. laws were primarily state laws. As the United States grew, so did the need for a more uniform law of the land. This chapter will discuss briefly some of the responsibilities of the state and local areas as compared with federal responsibilities. In addition, the chapter will go one step further to provide a few examples of how local regulations may impact you and the foods you eat. This overview comes from a compilation of articles from several government websites.

## National versus State Government

Our first type of government was based primarily on the state form of government. Prior to the signing of the Constitution, the United States of America had been made up of 13 colonies that had been ruled by England. Following the Revolutionary War, these colonies, although they had formed a loose alliance under the Articles of Confederation, governed themselves. They feared a strong central government such as that which governed them under England's rule. It was soon discovered, however, that this weak form of government could not survive, and thus the Constitution was drafted.[1] The Constitution

- defines and limits the power of the national government;

- defines the relationship between the national government and individual state governments; and
- guarantees the rights of the citizens of the United States.

It was decided that a system of government based on federalism would be established. In other words, power would be shared between the national and state (local) governments. The opposite of this system of government is a centralized government, such as that in France or England, where the national government maintains all power.

Sharing power between the national government and state governments allows U.S. citizens to enjoy the benefits of diversity and unity. For example, the national government may set a uniform currency system. Could you imagine having 50 different types of coins, each with a different value? You would need to take along a calculator to go shopping in another state. By setting up a national policy, the monetary system is fair to everyone, and the states do not have to bear the heavy burden of regulating their currency.

On the other hand, issues such as the death penalty have been left up to the individual states. The decision whether to have a death penalty depends on a state's history, needs, and philosophy.

## National Government

The delegates to the Constitutional Convention faced a difficult challenge. They wanted to ensure a strong, cohesive central government, yet they also wanted to ensure that no individual or small group in the government would become too powerful. Because of the colonies' experience under the British monarchy, the delegates wanted to avoid giving any one person or group absolute control in government. Under the Articles of Confederation, the government had lacked centralization, and the delegates didn't want to have that problem again. To solve these problems, the delegates to the Constitutional Convention created a government with three separate branches, each with its own distinct powers.[2] This system would establish a strong central government while ensuring a balance of power.

Governmental power and functions in the United States rest in three branches of government: the legislative, judicial, and executive. Article I of the Constitution defines the legislative branch and vests power to enact laws in the Congress of the United States. The executive powers of the president are defined in Article II. Article III places judicial power in the hands of one Supreme Court and inferior courts as Congress sees necessary to establish. A complete diagram of the branches of the U.S. government may be found in the *U.S. Government Manual* (from National Archives and Records Administration, Office of the *Federal Register*, Washington, DC).

In this system of separation of powers, each branch operates independently of the others; however, there are built-in checks and balances to prevent tyrannous concentration of power in any one branch and to protect the rights and liberties of citizens. For example, the president can veto bills approved by Congress, and the president nominates individuals to serve in the federal judiciary; the Supreme Court can declare a law enacted by Congress or an action by the president unconstitutional; and Congress can impeach the president and federal court justices and judges.

## State Government

State governments have their own constitutions, similar to that of the national Constitution; however, the laws made in individual states cannot conflict with the U.S. Constitution.[3] Each state's constitution is different because each state has its unique history, needs, philosophy, and geography.

During the first 100 years of United States history, the states did most of the governing that directly affected the people. The national government mainly concentrated on foreign affairs. This is known as *dual federalism*, where each level of government controlled its own sphere. During this time, however, a rift began to form between the state and federal governments over the issue of who had sovereignty, a rift that would culminate in the Civil War.

This issue was clarified after the Civil War, when a number of constitutional amendments were passed.

### The Civil War Amendments

The federal government's control over social and economic policy and protection of the civil rights of citizens were spelled out by a series of constitutional amendments, as follows.

- **Thirteenth Amendment:** "Neither slavery nor involuntary servitude, except as a punishment for crime . . . shall exist within the United States."
- **Fourteenth Amendment:** "All persons born or naturalized in the United States . . . are citizens of the United States . . . . No State shall make or enforce any law which shall abridge the privileges or immunities of citizens of the United States; nor shall any State deprive any person of life, liberty, or property, without due process of law; nor deny to any person within its justfication the equal protection of the laws."
- **Fifteenth Amendment:** "The rights of citizens of the United States to vote shall not be denied or abridged by the United States or by any State on account of race, color, or previous condition of servitude."

Since 1860, dual federalism continued, but the power of the federal government began to strengthen. The Great Depression in the 1930s brought the end of dual federalism because states were unable to cope with the economic upheaval. Instead, President Roosevelt's New Deal brought about a system of *cooperative federalism.* Instead of assigning specific functions to each level of government, Roosevelt encouraged the national, state, and local governments to work together on specific programs.

## Powers of the National Government and State Governments

The exclusive powers of the national government and state governments can be seen in Table 3.1. In addition to their exclusive powers, both the national government and state governments share powers.[4] Shared powers between the national government and state governments are called *concurrent powers.* Current powers of the national government and state governments include the ability to

- collect taxes,
- build roads,
- borrow money,
- establish courts,
- make and enforce laws,
- charter banks and corporations,
- spend money for the general welfare, and
- take private property for public purposes, with just compensation.

**Table 3.1.** Exclusive powers of the national government and state governments

| *National Government* | *State Governments* |
|---|---|
| • Print money | • Issue licenses |
| • Regulate interstate (between states) and international trade | • Regulate intrastate (within the state) businesses |
| • Make treaties and conduct foreign policy | • Conduct elections |
| • Declare war | • Establish local governments |
| • Provide an army and navy | • Ratify amendments to the Constitution |
| • Establish post offices | • Take measures for public health and safety |
| • Make laws necessary and proper to carry out these powers | • May exert powers the Constitution does not delegate to the national government or prohibit the states from using |

*Source:* Ben's Guide to U.S. Government for Kids. 2003. Exclusive Powers of the National Government and State Governments. Superintendent of Documents, U.S. Government Printing Office. Online http://bensguide.gpo.gov/9-12/government/federalism2.html (September 2004).

In addition, neither the national government nor state governments may[5]

- grant titles of nobility;
- permit slavery (13th Amendment);
- deny citizens the right to vote due to race, color, or previous servitude (15th Amendment); or
- deny citizens the right to vote because of gender (19th Amendment).

Those powers denied to the national and state governments can be seen in Table 3.2.

## Food-Related Laws and Regulations

In early years people grew their own food. As more people left the farm and moved to the cities, food was produced and processed farther away

**Table 3.2.** Powers denied to the national government and state governments

| *National Government* | *State Governments* |
|---|---|
| • May not violate the Bill of Rights | • May not enter into treaties with other countries |
| • May not impose export taxes among states | • May not print money |
| • May not use money from the treasury without the passage and approval of an appropriations bill | • May not tax imports or exports |
| • May not change state boundaries | • May not impair obligations of contracts |
| | • May not suspend a person's rights without due process |

*Source:* Ben's Guide to U.S. Government for Kids. 1999. Powers Denied the National Government and State Governments. Superintendent of Documents, U.S. Government Printing Office. Online http://bensguide.gpo.gov/9-12/government/federalism3.html (September 2004).

from where it was consumed. This distance and the multiple hands the food passed through brought about a need for laws and regulations to ensure the safety and quality of the food from farm to table. Many of these laws and regulations are imposed by state and/or local jurisdictions. The following are a few examples of food-related laws and regulations enforced by state and/or local jurisdictions.

## U.S. Department of Agriculture

The Food Safety and Inspection Service (FSIS), which is a part of the U.S. Department of Agriculture (USDA) is responsible for meat and poultry inspection. Federal inspection by FSIS of meat and poultry occurs in processing plants that ship products across state lines. If a processing plant sells its products in-state only, the inspection can be performed by state inspectors.

Establishments have the option to apply for federal or state inspection. Under the agreement, a state's program must enforce requirements "at least equal to" those imposed under the Federal Meat and Poultry Products Inspection Acts. However, products produced under state inspection are limited to intrastate commerce. FSIS provides up to 50 percent of the state's operating funds, as well as training and other assistance.

State Meat and Poultry Inspection (MPI) programs are an integral part of the nation's food safety system.[6] About 2,100 meat and poultry establishments are inspected under state MPI programs. All of these establishments are small or very small. State MPI programs are characterized as providing more personalized guidance to establishments in developing their food safety–oriented operations. FSIS provides approximately $43 million annually to support the 27 state MPI programs currently operating. The states participating in the MPI programs can be seen in Table 3.3. For states without state inspection, FSIS must provide for the inspection in the designated category regardless of whether the product is shipped via interstate commerce.

State MPI programs operate under a cooperative agreement with FSIS. Under the agreement, a state's program must enforce requirements "at least equal to" those imposed under the Federal Meat Inspection Act (FMIA) and the Poultry Products Inspection Act (PPIA). Comparable requirements for state inspection programs are defined in the FMIA for cattle, sheep, swine, goats, horses, mules and other equines and in the PPIA for domesticated poultry, defined by regulation to include chickens, turkeys, ducks, geese, guineas, ratites, or squabs. FSIS Cooperative Inspection Program functions are outlined below. The outline uses the authorities of the FMIA as an example. The PPIA has comparable provisions.

The secretary of agriculture is authorized under Title III, when it would effect the purposes of the FMIA, to cooperate with states in developing and administering an "at least equal to" program imposing mandatory requirements for

- antemortem, postmortem, reinspection, sanitation (Title I);
- denaturing product sold not for food (Title II Sec. 201);
- record maintenance (Title II Sec. 202);
- provision of access (Title II Sec. 202);
- registration of brokers, renderers, wholesalers, and others—including products for human food and dead, dying, disabled and diseased animals (Title II Sec. 203); and
- control of dead, dying, disabled, and diseased animals (Title II Sec. 204).

Cooperation includes

- advisory assistance in planning and developing the program,
- technical and laboratory assistance and training, and
- funding to 50 percent.

Cooperation is contingent upon administration of the state program in a manner in which the secretary, in consultation with advisory committee,

**Table 3.3.** List of states participating in the MPI program

| *State* | *Meat and/or Poultry Programs* |
|---|---|
| Alabama | Meat & poultry |
| Arizona | Meat & poultry |
| Delaware | Meat& poultry |
| Georgia | Meat only |
| Illinois | Meat & poultry |
| Indiana | Meat & poultry |
| Iowa | Meat & poultry |
| Kansas | Meat & poultry |
| Louisiana | Meat & poultry |
| Minnesota | Meat & poultry |
| Mississippi | Meat & poultry |
| Missouri | Meat & poultry |
| Montana | Meat & poultry |
| New Mexico | Meat & poultry |
| North Carolina | Meat & poultry |
| North Dakota | Meat only |
| Ohio | Meat & poultry |
| Oklahoma | Meat & poultry |
| South Carolina | Meat & poultry |
| South Dakota | Meat only |
| Texas | Meat & poultry |
| Utah | Meat & poultry |
| Vermont | Meat & poultry |
| Virginia | Meat & poultry |
| West Virginia | Meat & poultry |
| Wisconsin | Meat & poultry |
| Wyoming | Meat & poultry |

*Source:* Online
http://www.fsis.usda.gov/regulations/listing_of_participating_states/index.asp. (September 2004)

deems adequate to effect the purposes of the Act (Title III Sec. 301(a)(3) and (4). The state program is subject to designation (returned to federal inspection) if the state chooses to drop the program, or requirements defined in the FMIA are not met. The program may be designated per 9 *CFR* 331.2 covering Titles I and IV of the act (Title IV covers authorities for penalties, detention, seizures, and Federal Trade Commission Act provisions) and per 9 *CFR* 331.6 covering Title II of the act.

## The Food and Drug Administration

The role of the Food and Drug Administration (FDA) in providing assistance to state and local governments is derived from the Public Health Service Act (PL 78-410).[7] Responsibility for carrying out the provisions of the act relative to food protection was transferred to the Commissioner of Food and Drugs in 1968 (21 *CFR* 5.10(a)(2) and (4)).

Additionally, Sections 301(k) and 704 of the Federal Food, Drug, and Cosmetic Act provide authority to the FDA to inspect and take enforcement action with respect to retail food establishments where food is held after introduction into interstate commerce.

FDA–state alliances are governed by memoranda of understanding with various national conferences, where federal and state officials, along with industry representatives and others, meet either yearly or biannually to review or revise procedures. The states, who have voting authority, examine ordinances, operational manuals, or pertinent issues to make sure everyone is going to be using the same rules.

More than 3,000 state and local government agencies regulate the country's retail food industry. They are responsible for over one million food establishments: restaurants and grocery stores, as well as vending machines, cafeterias, and other outlets in health-care facilities, schools, and correctional facilities. The FDA's regional food specialists provide training, program evaluation, and technical assistance to these agencies.

### Shellfish Program

The shellfish program is governed by the Interstate Shellfish Sanitation Conference, which was established in 1982 and meets each summer for seven days. In April of each year, conference members—FDA, states, industry representatives, the Environmental Protection Agency, and the

National Marine Fisheries Service—submit ideas and issues for consideration by the conference. One of three task forces, made up of conference members, examines each issue and makes a recommendation to the general assembly, which includes only state representatives who vote on the issue.

FDA's Shellfish Sanitation Program has a regional staff that interacts with state officials to help bolster enforcement programs. These shellfish specialists audit state programs to ensure compliance with laws, regulations, and requirements of the National Shellfish Sanitation Program (NSSP) and other criteria agreed on by the Interstate Shellfish Sanitation Conference. Their goal is to ensure that state officials comply with the NSSP in classifying harvesting waters, controlling illegal harvesting, as well as overseeing processing plant conditions and product labeling.

The shellfish program also provides states with training, technical assistance, and results of scientific research.

### Milk Program

The FDA assists states in preventing disease transmitted through milk and helps enforce state milk regulations. It promotes and helps ensure compliance with the model Grade A Pasteurized Milk Ordinance, a document similar to the FDA's *Food Code*. FDA's regional milk specialists offer seminars to state officials to promote uniformity in interpreting the Pasteurized Milk Ordinance, as well as on other issues such as laboratory analysis methods.

The milk program has an auditing feature called check rating, which double-checks state evaluations of milk sources. The FDA produces a quarterly Interstate Milk Shippers list of ratings states have given to interstate milk plants and farms. The FDA then checks, or justifies, the rating the state has given.

### Retail Food Protection Program

The Retail Food Protection–State Program is a cooperative federal/state program concerned with the safety of food and the prevention and reduction of foodborne illness at the retail level. This program covers all aspects of retail food served or offered for human consumption at food establishments, which include, but are not limited to, restaurants, food stores, vending operations, and facilities serving highly susceptible populations (i.e., nursing homes and children's day-care centers). One example of this program is the FDA's *Food Code*.

*Food Code*

The *Food Code* includes standards for such things as cooking times and temperatures, refrigeration temperatures, and storage requirements for many kinds of foods.[8] In 1994, Rhode Island was the first state to implement the *Food Code* as law, but that was just a first step.

More than 3,000 state and local regulatory agencies assume primary responsibility for monitoring retail food operations and assure that industry is adequately protecting the consumer in the marketplace.

The FDA publishes the *Food Code*, a model that assists food control jurisdictions at all levels of government by providing them with a scientifically sound technical and legal basis for regulating the retail and food service segment of the industry. Local, state, tribal, and federal regulators use the FDA *Food Code* as a model to develop or update their own food safety rules and to be consistent with national food regulatory policy.

It also serves as a reference of best practices for the retail and food service industries (restaurants and grocery stores and institutions such as nursing homes) on how to prevent foodborne illness. Many of the over one million retail and food service establishments apply *Food Code* provisions to their own operations.

Between 1993 and 2001, the *Food Code* was issued in its current format every two years. With the support of the Conference for Food Protection, the FDA has decided to move to a four-year interval between complete *Food Code* revisions. The next complete revision of the *Food Code* will be published in 2005. During the four-year interim period, a *Food Code* supplement that updates, modifies, or clarifies certain provisions is being made available.

The changes contained in the supplement reflect the current science and emerging food-safety issues, and imminent health hazards related to food safety, allowing the most current food-safety provisions to be available for agencies planning to initiate rule-making activities prior to 2005. In addition, this supplement gives other users of the *Food Code*—such as educators, trainers, and the food service, retail food, and vending industries—up-to-date information regarding how to best mitigate risk factors that contribute to foodborne illness.

This supplement addresses recommendations made by the 2002 Conference for Food Protection with which the FDA, Centers for Disease

**Table 3.4.** Status of state and territorial *Food Code* adoptions

| *State or Territory* | *2000 Census* | *Retail Food Program Agency* | *Agency* Food Code *Status* | *Rule Making* |
|---|---|---|---|---|
| Alabama | 4,447,100 | Dept. of Public Health | Has adopted the *Food Code*, 1995 | Yes, 2004 |
| Alaska | 626,932 | Dept. of Environ. Conservation | Has adopted the *Food Code*, 1999 | Yes, 2004 |
| American Samoa | 57,291 | | Did not respond to survey | |
| Arizona | 5,130,632 | Dept. of Health Services | Has adopted the *Food Code*, 1999 | No |
| Arkansas | 2,673,400 | Dept. of Health | Based on 1976 Model Foodservice Code | Yes, 2004 |
| California | 33,871,648 | Dept. of Health Services | Has not adopted the *Food Code* | Yes, 2006 |
| Colorado | 4,301,261 | Dept. of Public Health and Environ. | Has adopted the *Food Code*, 1997 | Yes, 2004 |
| Connecticut | 3,405,565 | Dept. of Consumer Protection | Has adopted the *Food Code*, 1999 | Yes, 2004 |
| | | Dept. of Public Health | Has adopted the *Food Code*, 1997 | No |
| Delaware | 783,600 | Dept. of Health and Social Services | Has adopted the *Food Code*, 1999 | Yes, 2004 |
| District of Columbia | 572,059 | Dept. of Health | Has adopted the *Food Code*, 1999 | No |
| Florida | 15,982,378 | Dept. of Agriculture | Has adopted the *Food Code*, 2001 | No |
| | | Dept. of Bus and Prof. Regulation | Has adopted the *Food Code*, 1999 | Yes, 2004 |
| | | Dept. of Health | Has adopted the *Food Code*, 2001 | No |
| Georgia | 8,186,453 | Dept. of Agriculture | Has adopted the *Food Code*, 2001 | No |
| | | Dept. of Human Resources | Has adopted the *Food Code*, 1993 | Yes, 2004 |

**Table 3.4.** Status of state and territorial *Food Code* adoptions *(Continued)*

| *State or Territory* | *2000 Census* | *Retail Food Program Agency* | *Agency* Food Code *Status* | *Rule Making* |
|---|---|---|---|---|
| Guam | 154,805 | Dept. of Public Health | Has not adopted the *Food Code* | Unknown |
| Hawaii | 1,211,537 | Dept. of Health | Has adopted the *Food Code*, 1997 | No |
| Idaho | 1,293,953 | Dept. of Health and Welfare | Has adopted the *Food Code*, 1997 | Yes, 2005 |
| Illinois | 12,419,293 | Dept. of Public Health | Has adopted the *Food Code*, 1995 | Yes, 2005 |
| Indiana | 6,080,485 | Dept. of Health | Has adopted the *Food Code*, 1999 | Yes, 2004 |
| Iowa | 2,926,324 | Dept. of Inspections and Appeals | Has adopted the *Food Code*, 1997 | Yes, 2004 |
| Kansas | 2,688,418 | Dept. of Health and Environ. | Has adopted the *Food Code*, 1999 | Yes, 2004 |
| Kentucky | 4,041,769 | Dept. of Public Health | Based on 1976 Model Food Service Code | Yes, 2004 |
| Louisiana | 4,468,976 | Dept. of Health and Hospitals | Has adopted the *Food Code*, 1999 | No |
| Maine | 1,274,923 | Dept. of Agriculture | Has adopted the *Food Code*, 1999 | No |
| | | Dept. of Human Services | Has adopted the *Food Code*, 1999 | No |
| Maryland | 5,296,486 | Dept. of Health and Mental Hygiene | Based on 1976 Model Food Service Code | Yes, 2004 |
| Massachusetts | 6,349,097 | Dept. of Public Health | Has adopted the *Food Code*, 1999 | Yes, 2004 |
| Michigan | 9,938,444 | Dept. of Agriculture | Has adopted the *Food Code*, 1999 | No |
| Minnesota | 4,919,479 | Dept. of Agriculture | Has adopted the *Food Code*, 1995 | No |
| | | Dept. of Health | Has adopted the *Food Code*, 1995 | No |
| Mississippi | 2,844,658 | Dept. of Agriculture | Has adopted the *Food Code*, 2001 | No |
| | | Dept. of Health | Has adopted the *Food Code*, 2001 | No |

**Table 3.4.** Status of state and territorial *Food Code* adoptions *(Continued)*

| *State or Territory* | *2000 Census* | *Retail Food Program Agency* | *Agency* Food Code *Status* | *Rule Making* |
|---|---|---|---|---|
| Missouri | 5,595,211 | Dept. of Health | Has adopted the *Food Code*, 1999 | Yes, 2004 |
| Montana | 902,195 | Dept. of Health | Has adopted the *Food Code*, 1999 | Yes |
| Nebraska | 1,711,263 | Dept. of Agriculture | Has adopted the *Food Code*, 2001 | No |
| Nevada | 1,998,257 | Dept. of Health | Has adopted the *Food Code*, 1993 | No |
| New Hampshire | 1,235,786 | Dept. of Health and Welfare | Has adopted the *Food Code*, 2001 | No |
| New Jersey | 8,414,350 | Dept. of Health and Senior Services | Based on 1976 Model Food Service Code | Yes, 2005 |
| New Mexico | 1,819,046 | Environ. Dept. | Has not adopted the *Food Code* | No |
| New York | 18,976,457 | Dept. of Agriculture | Has adopted the *Food Code*, 1999 | No |
| | | Dept. of Health | Based on 1976 Model Food Service Code | Yes, 2005 |
| North Carolina | 8,049,313 | Dept. of Agriculture | Has adopted 21 *CFR*, Part 110 | No |
| | | Dept. of Environ. and National Resources | Based on 1976 Model Food Service Code | Yes, unknown |
| North Dakota | 642,200 | Dept. of Health | Has adopted the *Food Code*, 2001 | No |
| Northern Mariana Islands | 69,221 | | Did not respond to survey | Yes, 2004 |
| Ohio | 11,353,140 | Dept. of Agriculture | Has adopted the *Food Code*, 1999 | Yes, 2004 |
| | | Dept. of Health | Has adopted the *Food Code*, 1999 | Yes, 2005 |
| Oklahoma | 3,450,654 | Dept. of Health | Has adopted the *Food Code*, 1999 | Yes, 2005 |
| Oregon | 3,421,399 | Dept. of Agriculture | Has adopted the *Food Code*, 1999 | Yes, 2004 |
| | | Dept. of Health | Has adopted the *Food Code*, 1999 | Yes, 2005 |

**Table 3.4.** Status of state and territorial *Food Code* adoptions *(Continued)*

| *State or Territory* | *2000 Census* | *Retail Food Program Agency* | *Agency* Food Code *Status* | *Rule Making* |
|---|---|---|---|---|
| Pennsylvania | 12,281,054 | Dept. of Agriculture | Has adopted the *Food Code*, 2001 | No |
| Puerto Rico | 3,808,610 | Dept. of Health | Has adopted the *Food Code*, 2001 | No |
| Rhode Island | 1,048,319 | Dept. of Health | Has adopted the *Food Code*, 1993 | No |
| South Carolina | 4,012,012 | Dept. of Health | Has adopted the *Food Code*, 1993 | No |
| South Dakota | 754,844 | Dept. of Health | Has adopted the *Food Code*, 1995 | Yes, 2004 |
| Tennessee | 5,689,283 | Dept. of Agriculture | Based on 1982 Model Retail Food Store Code | No |
| | | Dept. of Health | Has adopted the *Food Code*, 1999 | No |
| Texas | 20,851,820 | Dept. of Health | Has adopted the *Food Code*, 1997 | Yes, 2004 |
| U.S. Virgin Islands | 108,612 | Dept. of Health | Has not adopted the Food Code | Yes, unknown |
| Utah | 2,233,169 | Dept. of Agriculture | Has adopted the *Food Code*, 1997 | Yes, 2005 |
| | | Dept. of Health | Has adopted the *Food Code*, 1999 | No |
| Vermont | 608,827 | Dept. of Health | Has adopted the *Food Code*, 2001 | No |
| Virginia | 7,078,515 | Dept. of Agriculture | Based on 1982 Model Retail Food Store Code | Yes, 2005 |
| | | Dept. of Health | Has adopted the *Food Code*, 1999 | Yes, 2004 |
| Washington | 5,894,121 | Dept. of Health | Based on 1976 Model Food Service Code | Yes, 2005 |
| West Virginia | 1,808,344 | Dept. of Health | Has adopted the *Food Code*, 1999 | No |
| Wisconsin | 5,363,675 | Dept. of Agriculture | Has adopted the *Food Code*, 1999 | Yes, 2004 |
| | | Dept. of Health | Has adopted the *Food Code*, 1999 | Yes, 2005 |
| Wyoming | 493,782 | Dept. of Agriculture | Has adopted the *Food Code*, 2001 | No |

*Source:* FDA. 2004. *Real Progress in Food Code Adoptions*. FDA Center for Food Safety and Applied Nutrition. Online http://www.cfsan.fda.gov/~ear/fcadopt.html.

Control and Prevention (CDC), and USDA concur. It carries the same weight as the *Food Code*.[9]

Forty-four of 56 states and territories (79 percent) have adopted codes patterned after the 1993, 1995, 1997, 1999, or 2001 versions of the *Food Code*. Those 44 states and territories represent 75 percent of the U.S. population.

Twenty-two of the 56 states and territories have adopted the 1999 *Food Code*, representing 37.5 percent of the U.S. population. Mississippi, Florida, Pennsylvania, Georgia, Wyoming, New Hampshire, Vermont, North Dakota, Nebraska, and Puerto Rico have adopted the 2001 *Food Code*, representing 16.7 percent of the U.S. population.

The status of state and territorial food code adoptions can be found in Table 3.4.

## Summary

State and local laws and regulations play an important role in our everyday lives. The USDA and FDA examples provided are just a few of the state and local regulations that impact the safety and quality of our food supply.

Meat and poultry establishments have the option to apply for federal or state inspection. Under the agreement with FSIS (USDA), a state's program must enforce requirements at least equal to those imposed under the Federal Meat and Poultry Products Inspection Acts. However, products produced under state inspection are limited to intrastate commerce. The FSIS provides up to 50 percent of the state's operating funds, as well as training and other assistance.

The FDA directs its activities toward the establishment of an effective system of state public health control programs by providing the leadership necessary to achieve nationwide implementation of uniform technical guidelines, administrative procedures, and regulatory standards.

## References

1. Ben's Guide to U.S. Government for Kids. 2000. National versus State Government. Superintendent of Documents, U.S. Government Printing Office. Online http://bensguide.gpo.gov/9-12/government/federalism.html (September 2004).

2. Ben's Guide to U.S. Government for Kids. 2003. Branches of Government. Superintendent of Document, U.S. Government Printing Office. Online http://bensguide.gpo.gov/9-12/government/branches.html (September 2004).
3. Ben's Guide to U.S. Government for Kids. 2000. State Government. Superintendent of Document, U.S. Government Printing Office. Online http://bensguide.gpo.gov/9-12/government/state/index.html (September 2004).
4. Ben's Guide to U.S. Government for Kids. 2003. Exclusive Powers of the National Government and State Governments. Superintendent of Document, U.S. Government Printing Office. Online http://bensguide.gpo.gov/9-12/government/federalism2.html (September 2004).
5. Ben's Guide to U.S. Government for Kids. 1999. Powers Denied the National Government and State Governments. Superintendent of Document, U.S. Government Printing Office. Online http://bensguide.gpo.gov/9-12/government/federalism3.html (September 2004).
6. USDA Food Safety and Inspection Service. 2004. State Inspection Programs. Online http://www.fsis.usda.gov/regulations/state_inspection_programs/index.asp (September 2004).
7. FDA. 2003. Federal/State Food Programs. Online http://www.cfsan.fda.gov/~dms/primecon.html (September 2004).
8. FDA. 2004. Real Progress in Food Code Adoptions. FDA Center for Food Safety and Applied Nutrition. Online http://www.cfsan.fda.gov/near/fcadopt.html.
9. FDA. FDA Food Code. Online http://www.cfsan.fda.gov/~dms/foodcode.html (Accessed August 25, 2004).

CHAPTER 4

# Major Laws and Regulations Related to Food Safety and Quality

Patricia Curtis, Auburn University
Wendy Dunlap

There are several laws and regulations that are associated with food safety and quality. This chapter discusses a few of the key laws and regulations on which we depend to ensure the safety of our food supply. The information in this chapter is a collection of information from a number of government websites. This information changes frequently, so check specific agency websites for the most current information.

## Food, Drug, and Cosmetic Act of 1938

The Food, Drug, and Cosmetic Act (FDCA)[1] is the primary food law in the United States. It protects the safety and quality of the food supply by prohibiting two acts; adulteration and misbranding. The FDCA also defines all terms used in the act, defines exactly what the prohibited acts are, and lays out the authority to enforce the law when it is broken. The FDCA can be found in Title 21 Chapter 9 of the *U.S. Code*. The law is divided into chapters and sections.

Sections of the FDCA are referred to in two ways, by FDCA section number and by *U.S. Code* section number. The FDCA, when passed by Congress, was divided into sections and chapters. The FDCA section numbers refer to these sections. Once the FDCA was codified, or "put into" the *U.S. Code*, the numbers changed, but the divisions basically stayed the same. For example, when a reference is made to Section 1 of the FDCA, it is referring to the Short Title. When a reference is made to 21 *U.S.C.* Sec. 301, it is also to the Short Title. The Federal Food, Drug, and Cosmetic Act (FDCA) is codified in Title 21-Food and Drugs, Chapter 9-Federal Food, Drug, and Cosmetic Act of the United States Code.

## Amendments to the FDCA

Since its passage in 1938, the FDCA has had many amendments. Below is a partial list of amendments to the FDCA. A complete listing of amendments, along with sections affected, may be found in Subchapter I-Short Title of the FDCA.

- Food Additives Amendment of 1958
- Color Additives Amendments of 1960
- Saccharin Study and Labeling Act (1977)
- Infant Formula Act of 1980
- Saccharin Study and Labeling Act Amendment of 1981
- Saccharin Study and Labeling Act Amendment of 1983
- Food and Drug Administration Act of 1988
- Nutrition Labeling and Education Act (NLEA) of 1990
- Dietary Supplement Act of 1992
- Nutrition Labeling and Education Act Amendment of 1993
- Dietary Supplement Health and Education Act of 1994

The acts prohibited by the FDCA are described in Chapter III, *USC* Section 331 (FDCA Sec. 301). The first thing the Prohibited Acts section of the FDCA does is prohibit adulteration and misbranding of food in interstate commerce. The FDCA applies only to food involved in interstate commerce, because it is a federal, not a state, law. State laws govern food being sold only within one state. Specifically, interstate commerce is defined as commerce between any state or territory and any place outside thereof, or commerce within the District of Columbia or within any other territory not organized with a legislative body. The FDCA takes care to prohibit adulteration and misbranding of foods in all stages of interstate commerce.

The FDCA also prohibits actions that hinder the Food and Drug Administration (FDA) in investigating adulterated or misbranded food, such as refusing to permit access to any records required by the FDCA, refusing to permit entry for inspection, and forging or falsely representing any official tag or mark authorized by the FDCA. Other acts that are prohibited by the FDCA include altering, mutilating, removing, etc., of the label of any food that has passed through interstate commerce if the food is for sale. This will result in the food being adulterated or misbranded. Also, offering oleomargarine that has been colored in a manner not consistent with the FDCA for sale is prohibited.

The conditions that render a food adulterated are described in 21 *U.S.C.* Section 342 (FDCA Section 402). In general, a food is adulterated if it contains a poisonous or deleterious substance, its economic value has been decreased without notifying the consumer, it contains an unsafe color additive, it is a confectionery that contains alcohol or a nonnutritive substance, or it is oleomargarine, butter, or margarine that is unfit for food. Each of these conditions is described below.

## If It Contains a Poisonous or Deleterious Substance . . .

In general, a food is considered adulterated if it contains a poisonous or deleterious substance; however, there are many poisonous or deleterious substances that cannot be avoided. There are also instances where the benefit of using the substance outweighs the potential harm caused by it. The writers of the act realized this and allowed for these substances in the manner described below.

- Poisonous or deleterious substances that are not added to the food, but are naturally occurring: The FDCA states that a food that contains a poisonous or deleterious substance that is not added will not be adulterated if the quantity of that substance does not ordinarily render it injurious to health. The FDA has established defect action levels to inform processors what levels are considered injurious to health.
- Added poisonous or deleterious substances that cannot be avoided or are required for processing: The law requires that the Secretary promulgate regulations that limit the amount of the substance allowed in the food before it is considered adulterated. This amount is based on how much of the substance is needed in the food, or what level cannot be avoided in the food. (This subsection does not apply to

pesticide chemicals on raw commodities, food additives, color additives, or new animal drugs, which are dealt with later in the statute.)
- Pesticide chemicals: Tolerances for pesticide chemicals are set by the Environmental Protection Agency (EPA). A raw agricultural commodity with levels of pesticide chemicals above the tolerance is adulterated.
- Food additives: Food additives are allowed only if they are considered by the FDA to be safe or, if unsafe, they must be used in the manner that is allowed by regulations. Any food with unsafe food additives present or food additives used illegally is adulterated.
- New animal drugs: New animal drugs must be approved by the FDA before use and must be labeled properly. Use of an unapproved new animal drug will cause the resulting food product to be adulterated. These provisions are covered in more detail in 21 *U.S.C.* Sec 360(b) (FDCA Sec 512).

Other instances where a poisonous or deleterious substance renders a food adulterated are listed below.

- The food contains any filthy, putrid, or decomposed substance, or it is unfit for food.
- The food was prepared, packed, or held under insanitary conditions where it could have been contaminated with filth.

To help processors and FDA investigators understand what constitutes "insanitary conditions", the FDA has Good Manufacturing Practice Regulations (GMPs). The first GMPs developed were "umbrella" GMPs and apply to all food processors. These regulations can be found in 21 *CFR* 110; they address the kinds of buildings, facilities, equipment, and materials needed; pest control; washing facilities; and cleaning of equipment. GMP regulations for specific commodities have also been published. The specific commodities include low-acid canned foods, acidified foods, and processing and bottling drinking water.

- The food comes from a diseased animal or from an animal that did not die by slaughter.
- The food is packed in a container that is composed of a poisonous or deleterious substance. For example, a food packaged in a container that contains poisonous polychlorinated biphenyls (PCBs) is considered adulterated because the PCBs could migrate into the food.
- The food is subjected to radiation and is not approved for radiation.

## Its Economic Value Has Been Decreased without Notifying the Consumer . . .

Even if a food does not contain a poisonous or deleterious substance, it can still be considered adulterated if any valuable constituents are left out of the food or if the food has been changed in a manner that results in a less valuable food. The FDCA lists these specific instances where this type of adulteration takes place.

- If any valuable constituent has been taken or left out of the product, it is considered adulterated.
- If any substance has been substituted, the food is adulterated.
- If the food is damaged or inferior, and this fact is covered up, the food is adulterated.

  In one case, white poppy seeds that had been colored with charcoal to appear as more expensive naturally dark seeds were deemed to be adulterated.
- If any substance is added to or packed with the food that increases its weight or bulk, reduces its quality or strength, or makes the food appear more valuable, than it is, it is adulterated.

These types of adulteration are considered economic adulteration and are addressed by many early food laws. Other examples of adulterated products include watered-down milk, pepper with husks and trash added to it, and bread with chalk added to make it appear white.

## It Contains an Unsafe Color Additive . . .

Foods are considered adulterated if they contain color additives that are unsafe under Section 721 of the FDCA (21 *U.S.C.* 379(e)). This section states that color additives must be approved for use by regulation and meet all criteria for use.

## It Is a Confectionery That Contains Alcohol or a Nonnutritive Substance . . .

Confectioneries are generally considered adulterated if they contain alcohol or a nonnutritive substance. Exceptions to this rule are listed below.

- If the confectionery contains a nonnutritive component that serves a functional purpose and is not considered harmful, it is not adulterated. A stick on a lollipop would serve a functional purpose, is typically not hazardous to health, and therefore would not render the lollipop adulterated.
- Confectionery that contains alcohol is not considered adulterated if the alcohol is from flavoring extracts and is present in an amount no greater than one-half of 1 percent by volume.
- Other nonnutritive components are permitted if they are needed for some practical purpose in the manufacture, packaging, or storage of the confectionery and if they do not deceive the consumer or render the food harmful.

## Oleomargarine, Butter, or Margarine

Any oleomargarine, butter, or margarine is adulterated if it consists of any filthy, putrid, or decomposed substance or if it is otherwise unfit for food.

The other major act prohibited by the FDCA is the act of misbranding. In general, a food is misbranded if the labeling of the food is false or misleading in any manner. This includes, but is not limited to, the following instances.

- The label is false and misleading.
- The food is sold under the name of another food.
- The food is an imitation food, but is not labeled as such.
- Its container is misleading.
- Its package does not have the required information.
- It contains artificial coloring, artificial flavoring, or a chemical preservative and is not labeled as such.
- The label is false and misleading.

This is the basic rule for determining if a food is misbranded. If none of the special cases described below apply, a food can still be considered misbranded on the basis that its label is false and misleading.

- The food is sold under the name of another food. An example is bread made from white flour being labeled as whole wheat bread, in which case it would be misbranded.
- The food is an imitation food but is not labeled as such. A food that attempts to mimic another food, but is not that food, must contain the

word "imitation" followed by the name of the food imitated. For example, one would expect strawberry cake frosting to contain strawberries. If strawberry flavoring is used instead, the frosting should be called "imitation strawberry frosting."

- Its container is misleading. Food that is packaged in a container that misleads the consumer is considered to be misbranded. This would include a small amount of food (two green beans) packaged in a much larger container than needed (a standard number-10 can).
- Its package does not have the required information. All food labels are required to have at the minimum

  1. the name and place of business of the manufacturer, packer, or distributor;
  2. an accurate statement of the quantity of the contents in terms of weight measure, or numerical count, unless otherwise excepted in the statute;
  3. the common or usual name of the food; and
  4. an ingredient list if two or more ingredients are used in the food.

- Required word(s) are not placed prominently on the label. If any word(s) that are required in the label by the FDCA are not placed prominently or conspicuously, the food could be deemed misbranded.
- The food identified on the label has a standard of identity, quality, or fill, or a definition that the product does not meet. Any food that has a standard or definition and fails to meet that standard or definition is misbranded unless that failure is stated on the label.
- The FDCA gives the FDA the authority to write food standards. These standards include standards of identity, standards of quality, and fill of container standards.
- A food contains artificial coloring, artificial flavoring, or a chemical preservative and is not labeled as such.
- Foods that contain artificial coloring, artificial flavoring, or a chemical preservative must bear labeling that states that fact unless exempted by regulations.

There are more special circumstances described in the act that will not be discussed here.

Another provision of the FDCA is that food standards and definitions may be promulgated whenever in the judgment of the secretary of agriculture such action will promote honesty and fair dealing in the interest of consumers (21 *U.S.C.* 341). Before the passage of the FDCA in 1938, there were no formal standards against which to measure products. Each time a product was challenged in court, the informal standards used to judge it had to be justified. In addition, many food industries were pushing for standards to keep cheap imitation products from being sold as the real thing. Food standards include standards of identity, standards of quality, and fill-of-container standards. Any product that purports to be a standardized food and does not meet the standards is considered misbranded.

## Standards of Identity

Standards of identity define what a given food product is, its name, and the ingredients that *must* be used, or may be used, in the manufacture of the food. Early food standards were written for traditional foods and were based on "time-honored recipes" that the average consumer was familiar with. They give the ingredients that must be in a particular product and often list optional ingredients that *may* be included in the product. Only foods that adhered to this recipe were allowed to be called that food name. In addition to ensuring that consumers knew what they were getting when they purchased a product, these standards also gave the FDA the ability to limit the addition of food additives to foods with food standards. For example, in the standard for fruit jams and preserves, found in 21 *CFR* Sec. 150.160, the definition is given as follows:

> the viscous or semi-solid foods, each of which is made from a mixture composed of one or a permitted combination of the fruit ingredients specified in paragraph (b) of this section and one or any combination of the optional ingredients specified in paragraph (c) of this section which meets the specifications in paragraph (d) of this section, and which is labeled in accordance with paragraph (e) of this section. Such mixture, with or without added water, is concentrated with or without heat. The volatile flavoring material from such mixture may be captured during concentration, separately concentrated, and added back to any such mixture, together with any concentrated essence accompanying any optional fruit ingredient.

It then goes on to state the optional ingredients.

> (c) The following safe and suitable optional ingredients may be used: (1) Nutritive carbohydrate sweeteners. (2) Spice. (3) Acidifying agents. (4) Pectin, in a quantity which reasonably compensates for deficiency, if any, of the natural pectin content of the fruit ingre-

dient. (5) Buffering agents. (6) Preservatives. (7) Antifoaming agents, except those derived from animal fat.

With the passage of the Food Additive Amendment of 1958, the use of food standards for this purpose became less important. In the 1960s this type of standard of identity came under attack from the food industry, which claimed that it did not allow innovation and new products. FDA took this into account in writing new food standards. Food standards written today are more flexible and require minimum amounts of ingredients of concern. For example, instead of listing the required ingredients for the breading on frozen breaded shrimp, the standard states that the breading should

> consist of suitable substances which are not food additives as defined in section 201(s) of the Federal Food, Drug, and Cosmetic Act; or if they are food additives as so defined, they are used in conformity with regulations established pursuant to section 409 of the act. Batter and breading ingredients that perform a useful function are regarded as suitable, except that artificial flavorings, artificial sweeteners, artificial colors, and chemical preservatives, other than those provided for in this paragraph, are not suitable ingredients of frozen raw breaded shrimp. Chemical preservatives that are suitable are:
>
> (1) Ascorbic acid, which may be used in a quantity sufficient to retard development of dark spots on the shrimp; and
> (2) The antioxidant preservatives listed in subpart D of part 182 of this chapter that may be used to retard development of rancidity of the fat content of the food, in amounts within the limits prescribed by that section.
>
> (21 *CFR* Sec. 161.175)

As society continues to demand healthier food with less fat and calories than traditional foods, food standards have had to be reevaluated. Under older regulations, foods such as ice cream required a minimum fat content or could not be called ice cream, but rather had to be ice milk. Foods with standards of identity are now allowed to be modified as long as they comply with 21 *CFR* 130.10. They must

1. comply with the provisions of the standard for the traditional standardized food except for the deviation described by the nutrient content claim;

2. not be nutritionally inferior to the traditional standardized food;
3. possess performance characteristics, such as physical properties, flavor characteristics, functional properties, and shelf life, that are similar to those of the traditional standardized food, unless the label bears a statement informing the consumer of a significant difference in performance characteristics that materially limits the use of the modified food (e.g., "not recommended for baking");
4. contain a significant amount of any mandatory ingredient required to be present in the traditional standardized food; and
5. contain the same ingredients as permitted in the standard for the traditional standardized food, except that ingredients may be used to improve texture, prevent syneresis, add flavor, extend shelf life, improve appearance, or add sweetness so that the modified food is not inferior in performance characteristics to the traditional standardized food.

It is important to note that before the NLEA of 1990, only the optional ingredients in a standard food had to be listed on the label. The rationale behind this was that consumers knew the basic ingredients of the food and needed to be alerted to the less common ones. Today, however, few people make such foods as jellies, ketchup, and juice in their home and know what ingredients are typically used. For that reason, the law now requires that all ingredients be listed on the label.

## Standards of Quality

Standards of quality are minimum standards only and establish specifications for quality requirements. For example, the quality standard for canned green beans sets a limit for fiber and prescribes the method to be followed to determine the fiber content.

## Fill-of-Container Standards

Fill-of-container standards define how full the container must be and how to measure it. There are a few foods that have fill-of-container standards, including certain fruits, vegetables, fish, shellfish, and nuts that are canned, packed in glass, or packed in semirigid containers. The fill-of-container regulations are found in 21 *CFR* Parts 130–169. According to requirements of laws and regulations enforced by the FDA, the existing fill-of-container standards for canned fruit and vegetables may be grouped as

1. those that require the maximum practicable quantity of the solid food that can be sealed in the container and processed by heat without crushing or breaking such component (limited to canned peaches, pears, apricots, and cherries).
2. those requiring a minimum quantity of the solid food in the container after processing. The quantity is commonly expressed either as a minimum drained weight for a given container size or as a percentage of the water capacity of the container.
3. those requiring that the food, including both solid and liquid packing medium, shall occupy not less than 90 percent of the total capacity of the container.
4. those requiring both a minimum drained weight and the 90 percent minimum fill.
5. those requiring a minimum volume of the solid component irrespective of the quantity of liquid (canned green peas and canned field peas). Fill-of-container standards specifying minimum net weight or minimum drained weight have been established for certain fish products.

The FDCA also contains definitions and requirements for specific foods, including dietary supplements, infant formulas, foods for special dietary use, oleomargarine, and bottled drinking water.

## Infant Formulas

Because infants are particularly susceptible to injury, the FDCA puts added requirements on the makers of infant formulas. An infant formula is "a food which purports to be or is represented for special dietary use solely as a food for infants by reason of its simulation of human milk or its suitability as a complete or partial substitute for human milk" (21 *U.S.C.* Sec 321(z)). The nutrients that must be present in an infant formula are listed in 21 *U.S.C.* Sec. 350a(i), along with the minimum level allowed. For example, all infant formulas must contain at least 1.8 grams of protein and 3.3 grams of fat.

All manufacturers of infant formula must begin with safe food ingredients, which are either generally recognized as safe (GRAS) or approved as food additives for use in infant formula. Once an infant formula product is formulated, current laws require that the manufacturer must provide FDA assurance of the nutritional quality of that particular formulation before marketing the infant formula. FDA has provisions that include requirements for certain labeling, nutrient content, manufacturers quality control procedures (to assure the nutrient content of infant formulas), as well as company records and reports. The FDA is also working to finalize a proposed rule for good manufacturing practice, quality control procedures, quality factors, notification requirements, and reports and records for the production of infant formulas.[2]

## Foods for Special Dietary Use

Section 403(j) of the FFDCA classes a food as misbranded: "If it purports to be or is represented for special dietary uses, unless its label bears such information concerning its vitamin, mineral, and other dietary properties as the Secretary of Health and Human Services determines to be, and by regulations prescribes as necessary in order to fully inform purchasers as to its value for such uses."[3]

Section 411(c)(3) of the FFDCA defines "special dietary use" as "a particular use for which a food purports or is represented to be used," including but not limited to

1. supplying a special dietary need that exists by reason of a physical, physiological, pathological, or other condition, including but not limited to the conditions of disease, convalescence, pregnancy, lactation, infancy, allergic hypersensitivity to food, underweight, overweight, or the need to control the intake of sodium; and
2. Supplying a vitamin, mineral, or other ingredient for use by humans to supplement the diet by increasing the total dietary intake; and
3. Supplying a special dietary need by reason of being a food for use as the sole item of the diet. Regulations (21 *CFR* 105) under this section of the federal FDCA prescribe appropriate information and statements that must be given on the labels of foods in this class.

Importers and foreign shippers should consult the regulations before importing foods represented by labeling or otherwise as foods for special dietary use. When foods for special dietary use are labeled with claims of disease prevention, treatment, mitigation, cure, or diagnosis, they must comply with the drug provisions of the federal FDCA, unless the claim is a health claim authorized by regulation.

## Oleomargarine

Due to lobbying by the dairy industry, the FDCA has a section that deals specifically with colored oleomargarine and margarine. Because colored margarine competes with butter, Congress wanted to make sure that it was not sold as butter. Regulations in 21 *U.S.C.* Sec. 347 state that

> no person shall sell, or offer for sale, colored oleomargarine or colored margarine unless
>
> (1) such oleomargarine or margarine is packaged,
> (2) the net weight of the contents of any package sold in a retail establishment is one pound or less,
> (3) there appears on the label of the package (A) the word "oleomargarine" or "margarine" in type or lettering at least as large as any other type or lettering on such label, and (B) a full and accurate statement of all the ingredients contained in such oleomargarine or margarine, and
> (4) each part of the contents of the package is contained in a wrapper which bears the word "oleomargarine" or "margarine" in type or lettering not smaller than 20-point type.

In addition, the law states that any restaurants serving colored margarine must make sure that customers are aware that they are eating margarine and not butter. The exact color of margarine is specified in the law as well.

## Bottled Drinking Water

In the United States, bottled water and tap water are regulated by two different agencies; the FDA regulates bottled water and the U.S. Environmental Protection Agency (EPA) regulates tap water (also referred to as municipal water or public drinking water). EPA's Office of Ground Water and Drinking Water has issued extensive regulations on the production, distribution, and quality of drinking water, including regulations on source water protection, operation of drinking water systems, contaminant levels, and reporting requirements.[5]

The FDA regulates bottled water as a food.[4] The FDCA provides the FDA with broad regulatory authority over food that is introduced or delivered for introduction into interstate commerce. Under the FDCA, manufacturers are responsible for producing safe, wholesome, and truthfully labeled food products, including bottled water products. It is a violation of the law to introduce into interstate commerce adulterated or misbranded products that violate the various provisions of the FDCA.

FDA has established specific regulations for bottled water in Title 21 of the *Code of Federal Regulations* (21 *CFR*), including standard of identity regulations (21 *CFR* § 165.110[a]) that define different types of bottled water, such as spring water and mineral water, and standard of quality regulations (21 *CFR* §165.110[b]) that establish allowable levels for contaminants (chemical, physical, microbial, and radiological) in bottled water. FDA also has established current good manufacturing practice (CGMP) regulations for the processing and bottling of bottled drinking water (21 *CFR* part 129). Labeling regulations (21 *CFR* part 101) and CGMP regulations (21 *CFR* part 110) for foods in general also apply to bottled water. It is worth noting that bottled water is one of the few foods for which FDA has developed specific CGMP regulations or such a detailed standard of quality.

### 21 *CFR* Part 129

These regulations require that bottled water be safe and that it be processed, bottled, held, and transported under sanitary conditions. Processing practices addressed in the CGMP regulations include protection of the water source from contamination, sanitation at the bottling facility, quality control to assure the bacteriological and chemical safety of the water, and sampling and testing of source water and the final product for microbiological, chemical, and radiological contaminants. Bottlers are required to maintain source approval and testing records to show to government inspectors. Checking adherence to part 129 regulations is an important part of FDA inspections of bottled water plants.

### 21 *CFR* § 165.110

This section establishes a standard of identity and a standard of quality for bottled water. Under the standard of identity (165.110[a]), FDA describes bottled water as water that is intended for human consumption and that is sealed in bottles or other containers with no added ingredients except that it may contain safe and suitable antimicrobial agents. Fluoride also may be added within the limits set by the FDA. The name of the food is “bottled water” or “drinking water.” The FDA also has defined various other types of bottled water, such as artesian water, artesian well water, ground water, mineral water, purified water, sparkling bottled water, and spring water (Table 4.1). Bottled water labeled with any of these terms must meet the appropriate definitions under the standard of identity or it will be considered misbranded under the FDCA. For example, a bottle labeled as containing mineral water must meet bottled water regulation and the following FDA criteria, among others: the water must contain no less than 250 parts per million (ppm) total dissolved solids; it must come from a geologically and physically protected underground water source; and it must contain no added minerals. "Mineral water" also must have a constant level and relative proportions of minerals and trace elements at the

**Table 4.1.** Various types of bottled water

| *Type* | *Definition* |
|---|---|
| Artesian water | Water from a well tapping a confined aquifer in which the water level stands at some height above the top of the aquifer. |
| Mineral water | Water that originates from a geologically and physically protected underground water source containing not less than 250 ppm total dissolved solids. Mineral water is characterized by constant levels and relative proportions of minerals and trace elements at the source. No minerals may be added to mineral water. |
| Purified water | Water that is produced by distillation, deionization, reverse osmosis. or other suitable processes and that meets the definition of "purified water" in the U.S. *Pharmacopeia*, 23d Rev., Jan. 1, 1995. As appropriate, it may also be called "demineralized water," "deionized water," "distilled water," or "reverse osmosis water." |
| Sparkling bottled water | Water that, after treatment and possible replacement of carbon dioxide, contains the same amount of carbon dioxide that it had at emergence from the source. |
| Springwater | Water derived from an underground formation from which water flows naturally to the surface of the earth at an identified location. Springwater may be collected at the spring or through a bore hole tapping the underground formation feeding the spring, but there are additional requirements for use of a bore hole. |

*Note:* For complete regulatory definitions, see 21 *CFR* 165.110(a)(2).
*Source:* FDA. 2002. Bottled Water Regulations and the FDA. Online http://www.cfsan.fda.gov/~dms/botwatr.html (September 2004).

point of emergence from the source, with due account being taken of natural fluctuation cycles. The FDA established its definitions for different types of bottled water in 1995 (60 *FR* 57076). These preempted state definitions existing at that time, some of which varied from state to state.

Under the standard of quality (165.110[b]), the FDA establishes allowable levels for contaminants in bottled water. There are microbiological standards that set allowable coliform levels; physical standards that set allowable levels for turbidity, color, and odor; and radiological standards that set levels for radium-226 and radium-228 activity, alpha-particle activity, and beta particle and photon radioactivity. The standard of quality also includes allowable levels for more than 70 different chemical contaminants. (For complete information on allowable levels for chemical or other contaminants, see 21 *CFR* 165.110[b].)

Section 165.110(b) also lists methods that the FDA will use to determine whether bottled water samples comply with the quality standard. Bottlers are not required to use these methods in their own facilities; alternate methods are acceptable. Whatever method they use, bottlers are responsible for ensuring that their bottled water can pass the tests used by FDA in its own laboratories, should testing be performed by the FDA.

What happens if bottled water contains a substance at a level greater than that allowed under the quality standard? Section 165.110(c) states that when the microbiological, physical, chemical, or radiological quality of bottled water is below that prescribed in the quality standard, the label of the bottled water bottle must contain a statement of substandard quality, such as "contains excessive bromate," "contains excessive bacteria," or "excessively radioactive." However, including a label of substandard quality may not be sufficient. Regardless of whether bottled water bears a statement of substandard quality, it is considered adulterated if it contains a substance at a level considered injurious to health under section 402(a)(1) of the FDCA.

Another noteworthy point about section 165.110 is that it allows the use of safe and suitable antimicrobial agents, such as ozone (see 21 *CFR* §184.1563 for details on ozone usage). The FDA does not specifically require that bottlers use antimicrobial agents in bottled water as long as the water is safe for human consumption.

## Inspections

One way the FDA ensures that food companies are complying with the FDCA is through inspection of factories, warehouses, or other establishments where food is stored. These establishments include those where food is manufactured, processed, packed, or held, and includes vehicles used to transport food.

Inspections are conducted for a variety of reasons, including

- routine scheduled inspections,
- surveillance, and
- in response to a complaint (complaints are received from various sources including consumers, other government agencies, and the trade).

Section 704 of the FDCA (21 *U.S.C.* 374) allows these inspections if they are conducted "at reasonable times and within reasonable limits and in a reasonable manner." Generally, a reasonable time would be during normal operating hours when management is present. Reasonable limits and a reasonable manner means that the facility should suffer minimal disruption during the inspection.

All plants should be prepared for an inspection *before* the inspection occurs. There should be an inspection plan, and a specific person should be designated to accompany the inspector during the inspection. It would be wise for that person to be familiar with the FDA's *Investigations Operations Manual*, which explains how to conduct an inspection and can give insight as to what the inspectors are told to look for.

## Recalls

The FDA may ask a firm to voluntarily recall product. FDA's "Guidance for Industry Product Recalls, Including Removals and Corrections" (http://www.fda.gov/ora/compliance_ref/recalls/ggp_recall.htm) describes this process. A recall is a firm's removal or correction of a marketed product that the FDA considers to be in violation of the laws it administers and against which FDA would initiate legal action; for example, seizure. During a recall, a firm can expect to work more closely with FDA than under almost any other circumstance. In fact, the first step, when a product must be recalled, is for the manufacturer or distributor to call the nearest FDA field office and talk with the recall coordinator.

FDA's main concerns during a recall are that the firm has determined the location of the product and organized the prompt removal from commerce of any suspect lots. FDA will then work with the firm to identify the cause of the problem and the corrections needed to prevent a recurrence.

Essentially, the procedures for a product recall are determined by the individual company; however, a proper recall system will include provisions for recordkeeping, handling product returns, liaison with the FDA, and public information. The efficiency of tracking and removing a product depends on how well records have been maintained throughout the production and distribution process.

The FDA has developed guidelines to aid companies with the recall process. As described in *FDA Recall Policies,* the guidelines categorize all recalls into one of three classes according to the level of hazard involved.

- Class I recalls are for dangerous or defective products that predictably could cause serious health problems or death. Examples of products that could fall into this category are a food found to contain botulinum toxin, a label mix-up on a lifesaving drug, or a defective artificial heart valve.
- Class II recalls are for products that might cause a temporary health problem or pose only a slight threat of a serious nature. One example is a drug that is under strength but that is not used to treat life-threatening situations.
- Class III recalls are for products that are unlikely to cause any adverse health reaction, but that violate FDA regulations. An example might be bottles of aspirin that contain 90 tablets instead of the 100 stated on the label.

## The Federal Meat Inspection Act

In 1906 Upton Sinclair published a book entitled *The Jungle*. This book described the insanitary conditions of meat-packing houses in Chicago and prompted an investigation of the meat industry. As a result, Congress passed the Federal Meat Inspection Act (FMIA),[6] one of the first federal consumer protection measures. This act

- established sanitary standards for slaughter and processing establishments;
- mandated *antemortem* inspection of animals (cattle, hogs, sheep, and goats);
- mandated *postmortem* inspection of every carcass; and
- required the continuous presence of government inspectors in all establishments that manufactured meat products for commerce.

Because the program depended heavily on veterinary skills, it was implemented by the USDA's Bureau of Animal Industry, which, during that first year, oversaw the inspection of nearly 50 million animals. It did not cover any poultry or poultry products.

## The Poultry Products Inspection Act

Before World War II, consumers purchased most poultry from small local farms. During the following years, poultry became increasingly available, and the market for dressed, ready-to-cook poultry expanded rapidly. This increase in commerce led to a new need to regulate poultry. As a result, the Poultry Products Inspection Act (PPIA)[7] was passed by Congress in 1957. It made inspection mandatory for all poultry products intended for distribution in interstate commerce. It was modeled after the FMIA.

## Wholesome Meat Act of 1967 and Wholesome Poultry Products Act of 1968

The Wholesome Meat Act of 1967[5] and the Wholesome Poultry Products Act of 1968[6] amended the FMIA and PPIA to assure uniformity in the regulation of products shipped in interstate, intrastate, and foreign commerce. These acts provide the statutory basis for the current meat and poultry inspection system.

To assure uniformity, these acts extended federal standards to intrastate operations, provided for state-federal cooperative inspection programs, and required that state inspection systems be "at least equal to" the federal system. Therefore, meat and poultry sold in *intrastate* commerce meet the same standards as that in interstate commerce. In addition, under these acts meat and poultry products from foreign countries that are sold in the United States must have been inspected under systems that are equivalent to that of USDA. System reviews are used to evaluate the laws, policies, and operation of the inspection systems in that country. This evaluation is conducted by the Foreign Programs Division of International Programs.

Both acts expanded USDA authority by providing stronger enforcement tools, including withdrawal or refusal of inspection services, detention, injunctions, and investigations. New regulatory authority over allied industries, including renderers, food brokers, animal food manufacturers, freezer storage concerns, transporters, retailers, and other entities were also granted.

These acts also incorporated adulteration and misbranding prohibitions similar to those in the FDCA relating to food and color additives, animal drugs, and pesticide chemicals.

## Acts Prohibited by FMIA Listed in 21 U.S.C. Sec. 610.

The acts prohibited by the PPIA can be found at 21 *U.S.C.* Sec. 458.

- It is illegal to slaughter animals or prepare them to be used as human food in a manner that is not in compliance with the requirements of the FMIA or PPIA.
- It is illegal to (1)sell, (2)transport, (3)offer for sale or transportation, or (4)receive for transportation, adulterated or misbranded food. It is also illegal to do any of the above with food that is required to be inspected but has not been inspected and passed.
- It is illegal to adulterate or misbrand any such articles which are capable of use as human food.

The FMIA also prohibits the slaughter or handling in connection with slaughter any such animals in any manner not in accordance with the Humane Methods of Slaughter Act of 1978 (7 *U.S.C.* Sec. 1901). This act amended the FMIA. FSIS has taken substantial and comprehensive action to ensure the humane treatment and slaughter of animals in establishments. In 2001, FSIS was able to hire district veterinary medical specialists to serve as the primary contact for humane handling and slaughter issues in each district. In 2003, FSIS issued a directive to provide FSIS inspection personnel additional information on humane handling verification procedures and to clarify enforcement actions to be taken for violations. In February 2004, FSIS implemented the electronic Humane Activities Tracking program (HAT) to document inspection activities that ensure livestock are humanely handled in federally inspected facilities. HAT provides FSIS with more accurate and readily available information on the activities and time spent by inspection personnel to ensure humane handling and slaughter requirements are met. The HMSA requires that humane methods for handling and slaughtering be used for all meat inspected by FSIS. This statute seeks to prevent needless suffering and results in safer and better working conditions. FSIS assigns inspectors to slaughter plants to ensure compliance with HMSA requirements for humane slaughter and handling methods.[8]

The PPIA prohibits the sale, transport, or receipt of slaughtered poultry from an official establishment that has not had the blood, feathers, feet, head, or viscera removed in accordance with the appropriate regulations.

## Official Devices, Marks, or Certificates

The FMIA and PPIA both require that all meat and poultry sold in interstate and foreign commerce must be inspected. FSIS ensures that only inspected products are sold by requiring that all inspected products have an official mark or legend[5,6] that states that it was "inspected and passed" in accordance with the FMIA or PPIA. Both acts also prohibit the following in regards to official devices, marks, or certificates.

- The forgery of any official device, mark, or certificate
- The use, alteration, detachment, defacement, or destroying of any official device, mark, or certificate without authorization from the Secretary of Agriculture
- Failure to use, detach, deface, or destroy any official device, mark, or certificate when prescribed by regulations
- The possession of any counterfeit, altered, or forged device, mark, or certificate, or of any product that bears a counterfeit, altered, or forged device, mark, or certificate
- False statements regarding any shippers statement or other official or nonofficial certificate
- Representation of an article as inspected and passed or exempt when actually it is not inspected and passed or exempt

Both laws also state the following (21 *U.S.C.* Sec. 461, Sec. 675):

> Any person who forcibly assaults, resists, opposes, impedes, intimidates, or interferes with any person while engaged in or on account of the performance of his official duties under this chapter shall be fined not more than $5,000 or imprisoned not more than three years, or both. Whoever, in the commission of any such acts, uses a deadly or dangerous weapon, shall be fined not more than $10,000 or imprisoned not more than ten years, or both. Whoever kills any person while engaged in or on account of the performance of his official duties under this chapter shall be punished as provided under sections 1111 and 1114 of Title 18.

## Adulteration and Misbranding

The FMIA and PPIA state that it is illegal to adulterate or misbrand meat or poultry products. In general, a meat or poultry product is adulterated if it bears or contains any poisonous or deleterious substance that may render it injurious to health. The specific instances that render a product adulterated are very similar to those outlined in the FDCA. In both the FMIA and the PPIA, the definition of "adulterated" appears in the definitions section.

## Standards of Identity

Both the FMIA and PPIA give the USDA the authority to set definitions and standards of identity, as well as standards of fill of container, as long as the standards established do not contradict any standards by FDA set under the authority of the FDCA.

Standards of identity state certain requirements that the food must meet in order to be called by a particular name. All standards of identity for meat products are located in 9 *CFR* Section 319. All standards of identity for poultry products are located in 9 *CFR* Section 381 Parts 381.155–381.174.

## Inspection

Both the FMIA and PPIA ensure that meat and poultry products will not be adulterated or misbranded by authorizing the USDA (and subsequently FSIS) to inspect the establishments.[5,6] Establishments must provide a work area for the inspectors and pay for any overtime that they work.

As stated in 21 *U.S.C.* Section 606 (FMIA), inspectors must examine and inspect all meat food products prepared for commerce in any slaughtering, meat-canning, salting, packing, rendering, or similar establishment. To do this, the inspectors shall have access at all times, by day or night, whether the establishment be operated or not, to every part of said establishment. All products found to not be adulterated shall be marked, stamped, tagged, or labeled as "inspected and passed." If a product is found to be adulterated, it must be marked, stamped, tagged, or labeled as "inspected and condemned." Any condemned meat products must be destroyed.

FSIS inspectors must

1. inspect animals before and after slaughter;
2. inspect the sanitary conditions in meat-processing establishments;
3. inspect products;
4. ensure the disposal of condemned or seized products; and
5. inspect all product labeling.

### Antemortem Inspection (Preslaughter)

Both the FMIA and the PPIA require that animals be inspected before they are slaughtered.[5,6] Each red meat animal is inspected at rest and in motion. Any animals that show symptoms of disease must be separated from the others and examined by a veterinary medical officer (VMO). If the animal is diseased, it is condemned by the VMO. Any animal that does not have specific symptoms, but is abnormal, is marked as "U.S. suspect" and is inspected closely after slaughter.

Chickens and turkeys are inspected as a group before slaughter on a flock or lot basis. Any birds that show symptoms of disease are condemned.

### Postmortem Inspection (Postslaughter)

After the animals are slaughtered, each carcass must be inspected by an FSIS inspector.[5,6] A VMO supervises this line inspection. For red meat, the inspector must examine the heads, viscera, and carcasses of each animal. For poultry, the viscera and carcasses are examined for all birds, along with the heads of older birds.

FSIS inspectors use *organoleptic* methods to examine the meat and poultry, including visual observations and *palpations*. For red meat, incision of tissues is also used. The inspectors are looking for any signs of disease during this inspection. The organs in particular are examined because they often show signs of disease. The presence of any surface contaminants such as fecal matter are not allowed. It is possible for an inspector to condemn only one part of the carcass and allow the other parts to be marked "U.S. inspected and passed."

Any meat or poultry that is deemed to be unfit for human consumption is marked as "U.S. retained." These carcasses are removed from the line and are examined by a veterinarian who will determine if the food is "U.S. inspected and condemned" or "U.S. inspected and passed."

### Alternate Poultry Inspection Systems

In addition to the traditional inspection just described, there are three alternate systems for the inspection of poultry.[8] These are the streamlined inspection system (SIS) and the new line speed (NELS) inspection system, both of which shall be used only for broilers and Cornish game hens, and the new turkey inspection (NTI) system, which shall be used only for turkeys.

The SIS system is used in high-speed poultry operations and increases inspector efficiency by requiring establishment employees, rather than inspectors, to perform tasks that control product quality instead of safety.

The NELS system and NTI system require the presence of a quality control program. Quality control programs are voluntary and require self-monitoring by the establishment to ensure that regulatory requirements are being met. FSIS inspectors then use the records generated from the establishment's self-monitoring system to determine if they are in compliance.

### Sanitation Inspection

Inspectors that are trained in the area of sanitation must inspect all regulated establishments to ensure that the sanitary conditions of the establishment are in compliance with any applicable rules and regulations.[9] If these rules and regulations are not being adhered to, the product is adulterated, and the inspector must refuse to allow said meat or meat food products to be labeled, marked, stamped or tagged as "inspected and passed."[10]

The general sanitation regulations are found in 9 *CFR* Sec. 416. These regulations cover such things as the condition of the building, rooms, floors, walls, ceilings, equipment, and the cleanliness and hygiene of official establishment personnel.

### Product Inspection

Processing establishments are treated a little differently from slaughter establishments. Processing establishments are those that make meat and poultry products such as sausage, deli meats, and frozen dinners. FSIS considers daily visits by inspectors adequate to meet the FMIA's and PPIA's requirements for "continuous inspection." This policy allows one inspector to inspect several plants every day, instead of staying in one plant the entire time it is operating. While at the establishment, inspectors examine plant sanitation, equipment, cooking procedures, facility layout, and product labels. They also check to be sure that any ingredients being used are handled and used properly. Inspectors are not required to inspect each individual meat or poultry product.

### Exemptions to FMIA and PPIA

The FMIA and PPIA exempt the following situations from the continuous inspection requirement.[5,6] The adulteration and misbranding provisions of the two laws are still applicable.

- The slaughtering by any person of poultry or meat animal of his own raising, and the processing by him and transportation in commerce of the products exclusively for use by him and members of his household, employees, and nonpaying guests is exempt.
- The custom slaughter of any poultry or meat animal for the owner for use by him and members of his household, employees, and nonpaying guests is exempt.
- The slaughter of poultry or meat animals as required by recognized religious dietary laws may be exempt.

There are other specific exemptions described in the FMIA and PPIA (21 *U.S.C* Sec. 623, 21 *U.S.C.* Sec. 464).

## Pathogen Reduction, Hazard Analysis, and Critical Control Point System Regulations

Traditionally, federal inspection has relied heavily on organoleptic methods such as sight, touch, and smell, to check for signs of illness or contamination of the animal. This is a reasonable method to screen out diseased animals. However, as meat and poultry breeding have become more sophisticated, the dangers from diseased animals have become less of a threat. Instead, meat and poultry processors are now concerned with a different hazard: pathogenic bacteria.

FSIS addressed this problem by asking the National Research Council (NRC) to evaluate the scientific foundation for its meat and poultry inspection system. This subsequent report recommended the use of hazard analysis critical control point (HACCP) systems to improve the safety of meat and poultry. In response to these petitions and to consumer outcry for safer meat and poultry products, the USDA published the Pathogen Reduction/HACCP final rule on July 25, 1996.

The Pathogen Reduction/HACCP rule was the first major change to meat and poultry inspection in 90 years. The rule has four elements:

- Sanitation standard operating procedures (SSOPs)
- HACCP plans
- Mandatory *Escherichia coli* testing in slaughter plants
- Pathogen reduction performance standards for *Salmonella*.

These elements have been implemented and enforced in addition to the carcass-by-carcass inspection.

## Sanitation Standard Operating Procedures

Under the Pathogen Reduction/HACCP rule, meat and poultry processors must develop and implement written SSOPs. It is important to remember that SSOPs must address *direct product contamination.* The traditional sanitation regulations still address any other sanitation issues.

These procedures, or the SSOP plan, must describe all procedures the establishment will conduct daily, before and during operations, sufficient to prevent direct product contamination or adulteration. These procedures must address how the food contact surfaces of the facility, equipment, and utensils are cleaned.

The SSOPs must also state how frequently the procedures are to be conducted and the employee(s) that will be responsible for the program. Daily records of the implementation and monitoring of the SSOPs must be maintained and made available to FSIS upon request.

FSIS inspectors must determine if the establishment is adhering to their program. If the inspector determines that either the SSOPs have not been implemented properly or that they are not preventing direct product contamination, the establishment must take corrective actions to assure that the problem will not occur again.

SSOPs were implemented in all plants on January 27, 1997. The complete regulations on SSOPs are located in 9 *CFR* Part 416.

## HACCP Plans

The HACCP concept was originally developed in the 1960s by NASA and The Pillsbury Co. as a means of ensuring that astronauts would not be stricken with foodborne illness while on a mission. It is a preventative control system that focuses on the elimination of hazards. The seven principles of HACCP are as follows:

1. Hazard analysis: determine what the hazards are and how to prevent them.
2. Critical control point identification: identify where the hazards can be controlled.
3. Establishment of critical limits: determine the critical limits for preventative measures.
4. Monitoring procedures: establish means of monitoring those points.
5. Corrective actions: determine a means of correction when the limits are exceeded.
6. Record keeping: establish effective documentation.
7. Verification and validation procedures: ensure a means for verification of the system and validation of the plan.

FSIS has written several generic HACCP models for plants to use as a guide. In addition, FSIS has a website devoted to presenting information on the Pathogen Reduction/HACCP regulations.

### Mandatory Escherichia coli Testing

To verify that the HACCP plans are adequate, plants are required to test carcasses for generic *Escherichia coli* as an indicator of fecal contamination. While not necessarily pathogenic, the presence of generic *E. coli* is an indicator of possible fecal contamination. Performance criteria are set for each commodity, and any product that fails to meet the criteria is considered out of compliance. The frequency of testing is based on the plant's production volume.

### Pathogen Reduction Performance Standards for Salmonella

The Pathogen Reduction/HACCP rule requires that raw ground products and chilled carcasses be tested for *Salmonella*, a leading cause of foodborne illness. These results must be below the baseline level determined by FSIS for each product. FSIS conducts the tests to determine the effectiveness of the establishment's HACCP plan.

### Significant Changes for Inspectors and Industry

This regulation drastically changed the way that FSIS inspectors and plant employees interact. For the past 90 years of federal inspection, inspectors have taken a direct "command and control" approach to regulation. The result of this method is that many plants have grown to depend on the inspector to tell them when they are out of compliance and how to correct the situation. FSIS has stated that this and the use of command and control techniques have often blurred the line between industry and regulator.

USDA intends for this new approach to clarify that industry, not the inspection service, is responsible for producing safe meat and poultry products. FSIS now sets performance standards, and each establishment may determine how they will meet those standards.

## References


1. *U.S. Code* Title 21 Food and Drugs, Chapter 9 Federal Food, Drug, and Cosmetic Act. Online http://www.access.gpo.gov/uscode/title21/chapter9.html (September 2004).
2. FDA. 2004. Infant Formula Overview. Online http://www.cfsan.fda.gov/%7Edms/inf-toc.html (September 2004).
3. FDA. What are FDA requirements for Foods for Special Dietary Use? Online http://www.cfsan.fda.gov/~dms/qa-ind6a.html (September 2004).
4. FDA. 2002. Bottled Water Regulations and the FDA. Online http://www.cfsan.fda.gov/~dms/botwatr.html (September 2004).
5. FSIS. Federal Meat Inspection Act. Online http://www.fsis.usda.gov/regulations_&_policies/Federal_Meat_Inspection_Act/index.asp (September 2004).
6. FSIS. Poultry Products Inspection Act. Online http://www.fsis.usda.gov/regulations_&_policies/Poultry_Products_Inspection_Act/index.asp (September 2004).
7. FSIS. 2004. FSIS Issues Notice of Humane Handling Requirements. News Release September 9, 2004, Online http://www.fsis.usda.gov/News_&_Events/NR_090904_01/index.asp (September 2004).
8. FSIS. 2004. FSIS Website. Online http://www.fsis.usda.gov/Home/index.asp (September 2004)
9. FSIS. 2003. Verifying an Establishment's Food Safety System. FSIS Directive 5000.1 Revision 1. Online http://www.fsis.usda.gov/OPPDE/rdad/FSISDirectives/5000.1Rev1.pdf (September 2004).
10. 9 *CFR* 416 Online http://www.access.gpo.gov/nara/cfr/waisidx_04/9cfr416_04.html (September 2004).

CHAPTER 5

# Food Labeling

Patricia Curtis, Auburn University
Wendy Dunlap

The Nutrition Labeling and Education Act (NLEA), which amended the Food, Drug, and Cosmetic Act (FDCA), requires most foods to bear nutrition labeling and requires food labels that bear nutrient content claims and certain health messages to comply with specific requirements. Although a number of regulations are discussed in this chapter, regulations are frequently changed. It is the responsibility of the food industry to remain current with the legal requirements for food labeling. All new regulations are published in the *Federal Register* prior to their effective date and compiled annually in Title 21 of the *Code of Federal Regulations*. Summaries of new regulations (proposed regulations and final regulations) are posted on the FDA's Internet website.[1]

## Food Labeling

The Food and Drug Administration (FDA) is responsible for ensuring that foods sold in the United States are safe, wholesome, and properly labeled. This applies to foods produced domestically, as well as foods from foreign countries. The FDCA and the Fair Packaging and Labeling Act (FPLA) are the federal laws governing food products under the FDA's jurisdiction.[1]

Under the FDA's laws and regulations, label approval is not required to import or distribute a food product.

Labeling of FDA-regulated foods is controlled under the authority of the FDCA as amended (21 *U.S.C.* Sec. 30) *et seq.* and the Fair Labeling and Packaging Act of 1966 (15 *U.S.C.* Sec. 1451) *et seq.*

The FDCA requires that the label of every processed, packaged food contains the name of the food, its net content, and the name and address of the manufacturer or distributor.

A list of ingredients also is required on most products. The law also prohibits statements in food labeling that are false or misleading.

The FPLA requires all consumer products in interstate commerce to contain accurate information and to facilitate value comparisons. The specific requirements of these laws are described in more detail below.

## Definitions

The FDCA defines *label* as a "display of written, printed, or graphic matter upon the immediate container of any article; and a requirement made by or under authority of this Act that any word, statement, or other information also appears on the outside container or wrapper, if any there be, of the retail package of such article, or is easily legible through the outside container or wrapper."

*Labeling* is defined as all labels and other written, printed, or graphic matter upon any article or any of its containers or wrappers, or accompanying such articles. Essentially, a label is on the actual package or container of food, and labeling is any accompanying information available at the place of purchase, as well as the label.

The purpose of food-labeling laws and regulations is to ensure that consumers are able to make informed decisions about a product based on its label and labeling. The FDCA and the FPLA, along with the resulting regulations, require that specific information be located on all food packages. Those requirements are described below.

Required information may either be placed all on one panel, the *principal display panel (PDP)*, or some specific information may be placed on the *information panel*. These two panels are described below.

## Principal Display Panel

The PDP is described both in the FPLA (15 *U.S.C.* 1459 (f)) and in the Food Labeling regulations (21 *CFR* Sec. 101.1).[2]

The PDP is the part of a label that is most likely to be displayed, presented, shown, or examined under customary conditions of display for retail sale. An example would be the front of a box of cereal.

If the package is likely to be displayed in more than one manner, an alternate PDP is required. This alternate PDP must contain the same information as the original PDP.

The size or area of the PDP is considered to be[2]

- in the case of a rectangular package, where one entire side properly can be considered to be the PDP side, the product of the height times the width of that side; and
- in the case of a cylindrical or nearly cylindrical container, 40 percent of the product of the height of the container times the circumference.

In the case of any containers shaped differently, 40 percent of the total surface of the container is considered to be the PDP, provided, however, that where such container presents an obvious PDP such as the top of a triangular or circular package of cheese, the area shall consist of the entire top surface.

In determining the area of the PDP, exclude tops, bottoms, flanges at tops and bottoms of cans, and shoulders and necks of bottles or jars. The statement of identity or name of the food and net quantity statement must appear on the PDP.

## Common or Usual Name

Regulations establishing common usual names for products can be found in 21 *CFR* 102. Some foods that have specific names are peanut spreads and onion rings made from dried onion. Fanciful names are product names that are not common names. For example, in the statement of identity Oh's Cereal, *Oh's* is the fanciful name and *cereal* is the common name.[3]

Where a food is marketed in various optional forms (whole, slices, diced, etc.), the particular form is considered a necessary part of the statement of identity and must be declared in letters of a type size bearing a reasonable relation to the size of letters forming the other components of the statement of identity. If the optional form is visible through the container or is depicted by an appropriate *vignette*, then the form does not need to be included in the statement.

A food that is a substitute for and resembles another food, but is nutritionally inferior to that food, is considered an imitation. Imitation foods must contain the word *imitation*, followed by the name of the food imitated in type of uniform size and prominence (21 *CFR* 101.3(e)).

Any beverage that purports to contain fruit juice must declare the percentage of juice present. This declaration should be near the top of the information panel and may be either "contains ____% juice" or "____% juice." The name of the fruit or vegetable may also be included (e.g., 100% apple juice).

Beverages that are described using the term *flavor* or *flavored,* that don't use the word *juice* other than on the ingredient statement, and that do not give the impression that they contain juice, are exempt from the percentage of juice requirement.

Only beverages that are 100 percent juice may use the word *juice* qualified with a term such as *beverage*, *drink*, or *cocktail*. The words *diluted ____ juice* may also be used. The rest of the juice regulations are in 21 *CFR* 101.30, 102, 33.

## Statement of Identity

Food labels are also required to have a statement of identity. This statement tells the consumers what they are buying. This statement must meet certain requirements regarding its appearance. A statement of identity must be[2]

- presented in bold type on the PDP,
- in a size reasonably related to the most prominent printed matter on such panel, and
- in lines generally parallel to the base on which the package rests as it is designed to be displayed.

The name used as the statement of identity must also conform to one of three standards:[2]

- If the FDA has published a definition and standard of identity for the product, that name must be the statement of identity;
- If there is no FDA definition or standard of identity for the product, it should be labeled using the common or usual name of that food; or
- If there is not a standard of identity or a common name, an appropriately descriptive term should be used. If the nature of the food is obvious, a fanciful name commonly used by the public for such food may be used.

Standards of identity define what a given food product is, its name, and the ingredients that *must* be used or may be used in the manufacture of the food. Early food standards were written for traditional foods and were based on time-honored recipes with which the average consumer was familiar. They give the ingredients that must be in a particular product and often list optional ingredients that *may* be included in the product. Only foods that adhered to this recipe were allowed to be called that food name. In addition to ensuring that consumers knew what they were getting when they purchased a product, these standards also gave the FDA the ability to limit the addition of food additives to foods with food standards.

It is important to note that before the NLEA of 1990, only the optional ingredients in a standard food had to be listed on the label. The rationale behind this was that consumers knew the basic ingredients and needed to be alerted to the less common ones. Today, however, few people make the foods with standards, such as jellies, ketchup, and juice in their home and know what ingredients are typically used. For that reason, the law now requires that *all* ingredients be listed on the label.

Food standards for specific foods are found in Sections 131–169 of Title 21 of the *Code of Federal Regulations*.

### Net Quantity of Contents

According to the FPLA and the FDCA, the net quantity of contents must appear on the PDP of packaged food.[4] The quantity given must be expressed in the appropriate terms of weight, measure, numerical count, or a combination of numerical count and weight or measure, in both the metric and U.S. customary system terms.

- If the food is liquid, the statement should be in terms of fluid measurement (U.S. gallon, quart, pint, or fluid ounce): for example, net contents 1 gal (3.79 L).
- If food is solid, semisolid, or viscous, or a mixture of solid and liquid, the statement should be in weight (pound and ounce): for example, net wt 1 lb 8 oz (680 g).
- If the food is a fresh fruit, fresh vegetable, or other dry commodity that is customarily sold by dry measure, the statement can be in terms of dry measure.
- If the food has a firmly established general consumer usage and trade custom of declaring the net quantity that is different from those described above, the established method may be used (ice cream is sold

in terms of liquid measure, gallons, even though it is more solid than liquid).

Quantities should be expressed at the appropriate temperature, also (frozen foods should be expressed at the frozen temperature).

It is important that the net quantity statement facilitates value comparisons and reduces consumer confusion. The term *net* refers to the amount of food that the consumers can be reasonably expected to get from the container. It does not apply to wrappers within the package or packing materials such as liquid that is not typically consumed.

The statement must appear in the lower 30 percent of the PDP in lines generally parallel to the base on which the package rests as it is designed to be displayed. If the package has a PDP of 5 square inches or less, then the statement may appear anywhere on the PDP. The statement should be separated from other printed information by at least a space equal to the height of the lettering on the top and bottom and by twice the width of the letter *N* on the left and right sides.

The declaration must appear in conspicuous and easily legible boldface print or type in distinct contrast to other matter on the package. Specific ratios for the height and width of the letters, as well as the type size requirements, can be found in 21 *CFR* Sec. 101.105 (h), (i).

The terms *net*, *net weight*, or *net contents* are not required, but are optional. Qualifying statements such as *jumbo quart* or *full pound* are prohibited in the net quantity statement.

### Information Panel

The information panel is described in 21 *CFR* Sec. 101.2.[2] When applied to packaged food, the information panel is that part of the label immediately contiguous and to the right of the PDP as observed by an individual facing the PDP with the following exceptions:[2]

- If the part of the label immediately contiguous and to the right of the PDP is too small to accommodate the necessary information or is otherwise unusable label space, for example, folded flaps or can ends, the panel immediately *contiguous* and to the right of this part of the label may be used.
- If the package has one or more alternate PDPs, the information panel is immediately contiguous and to the right of any PDP.
- If the top of the container is the PDP, and the package has no alternate PDP, the information panel is any panel adjacent to the PDP.

The term *information panel labeling* refers to label statements that are usually found on the information panel. These statements are the name and address of the manufacturer, packer, or distributor; the ingredient list; and nutrition label; and these elements must be placed together.

## Statement of Ingredients

Any food that is composed of two or more ingredients must declare those ingredients by common or usual name in descending order of predominance by weight on either the PDP or information panel (21 *CFR* Sec.101.4).[2] This means that the ingredients that are present in the highest amounts must be listed before those present in the lowest amounts. The statement must be on the same panel as the manufacturer's name and address.

### Ingredients Present in Amounts of 2 Percent or Less by Weight

Ingredients present in amounts of 2 percent or less by weight do not have to be listed in descending order of predominance.[2] Instead, they may be listed at the end of the ingredient statement using the following quantifying statement, "Contains 2 percent or less of *Ingredient*," or "less than 2 percent of *Ingredient*." If the ingredients are present in amounts lower than 2 percent, then the appropriate levels of 1.5 percent, 1.0 percent, or 0.5 percent may be used instead. No ingredient to which the quantifying phrase applies may be present in an amount greater than the stated threshold.

### Use of Collective or Generic Ingredient Names

The regulations state that "the name of an ingredient shall be a specific name and not a collective (generic) name" except in a few specific cases. For ingredients that are actually made of more than one ingredient, the common name of the ingredient may be given, followed by a parenthetical listing of all ingredients contained therein in descending order of predominance. For example, milk chocolate chips in a chocolate chip cookie may be listed in the statement as

> INGREDIENTS: . . . Milk chocolate chips (milk chocolate contains sugar; milk; cocoa butter; chocolate; soya lecithin, an emulsifier; and vanillin, an artificial flavoring) . . . .

See 21 *CFR* Section 101.4 for more details on listing ingredients. Foods with incorrect ingredient statements are considered to be misbranded.

### Foods with a Standard of Identity

In the past, producers of standardized foods were not required to declare the product ingredients on the label. This was because standardized foods were foods that were commonly prepared in the home, so consumers knew the ingredients that were typically found in those products. Now that fewer consumers prepare those foods in their homes, the ingredients present in those foods are not commonly known. For that reason, ingredient statements are now required for standardized foods.

### Allergens

Listing all ingredients present in a food is very important to people who have food allergies. A very small amount of a food can be deadly to a person who is allergic to that food.

### FDA Views on Allergens

FDA believes there is scientific consensus that the following foods (peanuts, soybeans, milk, eggs, fish, crustacean, tree nuts, and wheat) can cause serious allergic reactions in some individuals and account for more than 90 percent of all food allergies.[5]

Manufacturers are responsible for ensuring that food is not adulterated or misbranded as a result of the presence of undeclared allergens. Therefore, the FDA inspectors pay particular attention to situations where these substances are added intentionally to food, but not declared on the label, or may be unintentionally introduced into a food product and consequently not declared on the label. When an allergen not formulated in the product is identified as likely to occur in the food due to the firm's practices, (e.g., use of common equipment, production scheduling, rework practices), then the FDA inspector determines if a manufacturer has identified and implemented control(s) to prevent potential allergen cross-contact, for example, dedicated equipment, separation, production scheduling, sanitation, proper rework usage (like into like).[5]

Products that contain an allergenic ingredient by design must comply with section 403(i)(2) of the federal FDCA, which requires each ingredient in a food to be declared. Processing aids that contain allergenic ingredients must be declared in accordance with 21 *CFR* 101.4(a)(1).

Production practices that lead to unintentional addition of allergens to food may be considered insanitary conditions that may render the food injurious to health and cause the food product to be adulterated under section 402(a)(4) of the FDCA.[6]

The only exemption to labeling requirements is found in section 403(i)(2) of the FDCA and provides that spices, flavors, and certain colors used in food may be declared collectively without naming each. In some instances, these ingredients contain subcomponents that are allergens; therefore, the FDA strongly encourages the declaration of any allergenic ingredient contained in a spice, flavor, or color.[6]

### Food Safety and Inspection Service Views on Allergens

The Food Safety and Inspection Service (FSIS) recognizes that there are many foods and food ingredients to which some individuals may have some degree of intolerance or possible allergic reaction, which is why complete labeling is so critically important.[7] The FSIS supports practices that promote accurate informative product labeling, including voluntary statements on labels that alert people who have sensitivities or intolerances to the presence of specific ingredients. For example, a phrase such as "Contains: milk, wheat gluten, soy," has been accepted by the FSIS on labeling immediately following the ingredients statement. Additionally, further clarification of the source of a specific ingredient in a parenthetical statement in the ingredients statement on labeling, for example, "whey (from milk)," is encouraged as a means of informing consumers who may be alerted to a more recognizable term.

In limited situations, the use of factual labeling statements about a product's manufacturing environment, for example, "Produced in a plant that uses peanuts," may be used where good manufacturing practices and effective sanitation standard operating procedures (SSOPs) cannot reasonably eliminate the unintended presence of certain ingredients. For example, where chopped peanuts are used in making a dry Thai-style meat sauce mix, the necessity exists for a dry processing environment, and, thus, the production equipment cannot be washed with water or other fluids. In this instance, peanut dust may become airborne and unavoidably contaminate other meat or poultry products manufactured in the same production area. In this situation, a statement about the manufacturing environment, as described above, or the use of a "may contain (name of allergenic ingredient)" statement may be used on meat and poultry product labeling. This type of statement is not acceptable, however, where it is used as a replacement for poor SSOPs, that is, cross-contact because of an establishment's failure to adequately wash equipment between the manufacture of different lines of products. The FSIS believes the indiscriminate use of such elective statements does not promote good manufacturing practices under a hazard analysis critical control point system and is not helpful for consumers. Consequently, the use of factual statements about a product's manufacturing environment, for example, "Produced in a plant that uses peanuts," and the use of *may contain* statements, for example, "may contain peanuts," may be used only in cases where establishments show that adequate SSOPS cannot effectively eliminate the cross-contact issue.

The FSIS will consider any non-misleading symbols, statements, or logos that industry may want to include on labeling to inform consumers of the presence of potential food allergens in meat, poultry, or egg products. Requests to consider such features need to be submitted to the FSIS as a policy inquiry and not as label-approval submissions.

## Name and Address of Manufacturer, Packer, or Distributor

The label of a food in packaged form must specify conspicuously the name and place of business of the manufacturer, packer, or distributor. The actual name of the business, or the corporate name and the particular division of the corporation, should be used. If the food is not manufactured by the person whose name is on the label, the name shall be qualified by a phrase such as "Manufactured for (name of business)" or "Distributed by (name of business)." The place of business should include the street address, city, state, and zip code, unless the street address may be found in a current city directory or telephone directory. This information is usually printed on the information panel. More information may be found in 21 *CFR* Sec. 101.5.

## Country of Origin

Every article of foreign origin (including food) imported into the United States must be marked in a conspicuous place to indicate to the ultimate purchaser in the United States the English name of the country of origin of the article (19 *U.S.C.* Sec. 1304(a)).

The FDA *country of production* is defined in § 1.276(b)(4) of the Prior Notice in the *Federal Register*. For food that is in its natural state (21 *CFR* 1.276(b)(4)(i)), the FDA considers the country of production generally to be the country where the food was grown or collected, including harvested and readied for shipment to the United States. Articles of food grown, including harvested or collected and readied for shipment, in U.S. territories are considered to be grown in the United States (21 *CFR* 1.276(b)(4)(i)). For wild fish, however, including seafood, that is caught or harvested outside U.S. waters by a vessel that is not registered in the United States, the FDA's country of production is the country in which the vessel is registered. (21 *CFR* 1.276(b)(4)(i)).[8]

For food that is no longer in its natural state, the FDA's country of production is generally the country where the food was made or processed. If the article is made from wild fish aboard a vessel, however, the FDA country of production is the country in which the vessel is registered. If food that is no longer in its natural state was made in a U.S. territory, the FDA country of production is the United States (21 *CFR* 1.276(b)(4)(ii)).[8]

The FDA's country of production may be different from the Customs and Border Protection's definition of the country of origin. For example, the Customs and Border Protection's country of origin for beans that are grown and dried in the United States, then rehydrated and canned in the Dominican Republic, would be the United States. The FDA would consider the country of production to be the Dominican Republic. For purposes of the prior notice provisions of the FDCA, however, the article of food is canned beans, not dried beans. From a food safety standpoint, the FDA is most interested in knowing where the article of food was processed and canned. To avoid confusion between FDA's prior notice requirements and Custom and Border Protection's requirements, the interim final rule uses the term *FDA country of production* instead of the term *originating country* or *country from which the article originates*.[8]

All meat and poultry products imported into the United States are required to bear the country of origin on the labeling of the container in which the products are shipped, as well as the establishment number assigned by the foreign meat inspection system and certified to the U.S. Department of Agriculture (USDA). If imported meat or meat products are intended to be sold intact to a processor, wholesaler, food service institution, grocer, or the household consumer, that is, in the packaging and with the labeling on the package that enters the United States, the country of origin labeling (as well as the foreign establishment number) is conveyed to those market segments and recipients. For example, a bulk-shipping container that is imported from Canada, which consists of prepackaged and labeled meat cuts that are intended to be sold to institutional food service, grocers, or at retail to household consumers as they are packaged, would each bear country of origin labeling (e.g., Product of Canada) to the point the cuts reach consumers. Another example is a canned ham imported from Denmark that is sold intact; the labeling of such a product would have to bear the country of origin, Product of Denmark.[9]

If the imported meat or meat products are further processed in the United States, that is, removed from their containers and packaging and cut up or processed in any way, country of origin labeling is no longer required on the newly produced products or subsequent products made from them. For example, if canned hams from Denmark (labeled as Product of Denmark on entry to the United States) are shipped to a manufacturing establishment in the United States and made into ham salad, the country of origin of the ham is not required on the labeling of the ham salad product.[9]

### Artwork

Any artwork used on the label must not be misleading. In addition, the artwork may not hide or detract from any required labeling.

## Nutritional Labeling

In 1990, Congress passed the Nutrition Labeling and Education Act (NLEA), which amended the federal FDCA. This act and the resulting regulations require nutrition labeling for virtually all regulated foods. The goals of the act and regulations were to make nutritional information practically universal in the marketplace, show how each food fits into an overall healthy diet, and allow credible marketing claims. Due to the NLEA, about 90 percent of processed food carries nutrition information. Failing to adhere to the requirements of the NLEA will result in that food being misbranded. This act amended the federal FDCA of 1938. The majority of the requirements of this act can be found in 21 *U.S.C.* Sec. 343(q) and (r). The regulations supporting the NLEA are located in 21 *CFR* Sec 101.9.

## Nutrition Facts

The NLEA requires that all nutrition information be contained in a Nutrition Facts box. The regulations are very specific about the type size, font, and thickness of the lines used in the Nutrition Facts box.

The Nutrition Facts label may be placed on the PDP or the information panel. The information provided must all appear on the same panel and may not be split between panels.

**Nutrition Facts**

Serving size 1 cup (59g)
Servings Per Container about 7

| Amount Per Serving | Cereal | Cereal with 1/2 cup Fat Free Milk |
|---|---|---|
| **Calories** | 240 | 280 |
| Calories from Fat | 50 | 50 |
| | **% Daily Value **** | |
| **Total Fat** 6g* | **9%** | **9%** |
| Saturated Fat 0.5g | **3%** | **3%** |
| **Cholesterol** 0mg | **0%** | **0%** |
| **Sodium** 250 mg | **10%** | **13%** |
| **Potassium** 200mg | **6%** | **11%** |
| **Total Carbohydrate** 44g | **15%** | **17%** |
| Dietary Fiber 5g | **20%** | **20%** |
| Sugars 11g | | |
| Other Carbohydrate 28g | | |
| **Protein** 5g | | |
| Vitamin A | 15% | 20% |
| Vitamin C | 0% | 0% |
| Calcium | 0% | 15% |
| Iron | 90% | 90% |
| Vitamin D | 10% | 25% |
| Thiamin | 25% | 30% |
| Riboflavin | 25% | 35% |
| Niacin | 25% | 25% |
| Vitamin $B_6$ | 25% | 25% |
| Folic Acid | 50% | 50% |
| Vitamin $B_{12}$ | 25% | 35% |
| Phosphorus | 15% | 25% |
| Magnesium | 15% | 20% |
| Zinc | 10% | 15% |
| Copper | 8% | 8% |

*Amount in Cereal. One half cup fat free milk contributes an additional 40 calories, 65mg sodium, 200mg potassium, 6g total carbohydrate (6g sugars), and 4g protein.

**Percent Daily Values are based on a 2,000 calorie diet. Your daily values may be higher of lower depending on you calorie needs:

| | Calories | 2,000 | 2,500 |
|---|---|---|---|
| Total Fat | Less than | 65g | 80g |
| Saturated Fat | Less than | 20g | 25g |
| Cholesterol | Less than | 300mg | 300mg |
| Sodium | Less than | 2,400mg | 2,400mg |
| Potassium | | 3,500mg | 3,500mg |
| Total Carbohydrate | | 300g | 375g |
| Dietary Fiber | | 25g | 30g |

### Serving Size (or Unit of Measurement) (21 CFR Sec. 101.9(d)(3))

Serving size information must immediately follow the title Nutrition Facts. The serving size and the servings per container must both be listed. The amount per serving must be presented as a subheading. Serving sizes should be the amount that a typical consumer would eat at one sitting. They are based on FDA-established lists of "Reference Amounts Customarily Consumed per Eating Occasion" which is located in 21 *CFR* Sec. 101.12. This is to ensure uniformity that allows consumers to more easily compare the nutritional qualities of similar products.

Serving sizes must be given in both common household measures and metric measures (grams and milliliters). Common household measures include the cup, tablespoon, teaspoon, piece, slice, fraction (such as ¼ pizza), and common household containers used to package food (such as a jar or tray). Ounces may be used only if a household unit is not applicable and an appropriate visual is given: for example, 1 oz (28 g/about ½ pickle).

## Calorie Information (21 CFR Sec. 101.9(d)(5))

The number of calories per serving, as well as the calories from fat, must be listed on one line immediately following the amount per serving heading. A statement of the caloric content per serving is listed, expressed to the nearest 5-calorie increment up to and including 50 calories, and 10-calorie increments above 50 calories, except that amounts less than 5 calories may be expressed as zero. A statement of the caloric content derived from total fat in a serving, expressed to the nearest 5-calorie increment, up to and including 50 calories, and the nearest 10-calorie increment above 50 calories, should be listed. Label declaration of calories from fat is not required for products that contain less than 0.5 gram of fat in a serving, and amounts less than 5 calories may be expressed as zero. If calories from fat is not required and, as a result, not declared, the statement "Not a significant source of calories from fat" shall be placed at the bottom of the table of nutrient values in the same type size.

A statement of the caloric content derived from saturated fat in a serving may be declared voluntarily, expressed to the nearest 5-calorie increment, up to and including 50 calories, and the nearest 10-calorie increment above 50 calories, except that amounts less than 5 calories may be expressed as zero. This statement shall be indented under the statement of calories from fat.

On July 11, 2003, the FDA published a final rule in the *Federal Register* that amended its regulations on food labeling to require that *trans* fatty acids be declared in the nutrition label of conventional foods and dietary supplements (68 *FR* 41434). This rule is effective January 1, 2006.

The listing of *trans* fatty acids is mandatory even when mono- and polyunsaturated fatty acids are not listed. *Trans* fatty acids should be listed as "*Trans* fat" or "*Trans*" on a separate line under the listing of saturated fat in the nutrition label. *Trans* fat content must be expressed as grams per serving to the nearest 0.5-gram increment below 5 grams and to the nearest gram above 5 grams. If a serving contains less than 0.5 gram, the content, when declared, must be expressed as "0 g." For conventional food products (those food products other than dietary supplements), declaration of 0 g of *trans* fat is not required for such products that contain less than 0.5 g of total fat in a serving if no claims are made about fat, fatty acid, or cholesterol content. In the absence of these claims, the statement "Not a significant source of *trans* fat" may be placed at the bottom of the table of nutrient values in lieu of declaring 0 g of *trans* fat.

The labeling of dietary supplements is different from the labeling of conventional foods. Certain nutrients in conventional foods, when not present or when present at levels that the FDA has determined to be zero (see 21 *CFR* 101.9(c)), must be listed as zero on conventional food labels. However, when those same nutrients are not present in dietary supplements or present in dietary supplements at levels that the FDA has determined for conventional foods to be zero, such nutrients must not be listed on dietary supplement labels. An amount of 0 g and the "Not a significant source . . ." statement are not allowed in the nutrition labeling of dietary supplements (i.e., supplement facts). Consequently, when the amount of *trans* fat in a dietary supplement is less than 0.5 gram per serving, *trans* fat must not be listed on the Supplement Facts panel.[10]

## Nutrient Declarations (21 CFR Sec. 101.9(c)(d))

The amount of each nutrient by weight in grams as well as the percent daily value (DV) must appear for the nutrients as is seen in Table 5.1. A list of mandatory and optional nutrients can be seen in Table 5.1. The percent daily value must appear in boldface type on the nutrition label. Other nutrients are optional.

When a nutrient is added as a supplement or a claim is made about the nutrient on the label, that nutrient becomes mandatory. For example,

**Table 5.1.** Nutrient declarations on nutrition label

| *Mandatory* | *Voluntary* |
|---|---|
| Total fat | Polyunsaturated fat |
| Saturated fat* | Monosaturated fat |
| Cholesterol** | Potassium |
| Sodium | Dietary fiber |
| Total carbohydrates | Soluble fiber |
| Sugar*** | Insoluble fiber |
| Protein | Sugar alcohol |
| Vitamin A | Other carbohydrate |
| Vitamin C | |
| Calcium | |
| Iron | |

*Not required for products containing less than 0.5 gram of total fat per serving if no claims are made about fat or cholesterol content.
**Not required for products containing less than 2 milligrams of cholesterol per serving and making no claims about fat, fatty acids, or cholesterol.
***Not required for products containing less than 1 gram per serving unless claim is made about sweeteners, sugars, or sugar alcohol content.

declaration of folic acid content is voluntary, but if a label declares that a food is "a good source of folic acid," the percent DV of folic acid per serving must be on the label.

There are also requirements on how close the amount of nutrients must be to the amount stated on the label. In 21 *CFR* Sec. 101.9(g) it states that added nutrients must be present in amounts at least equal to the value for that nutrient declared on the label. Naturally occurring nutrients must be present in amounts at least equal to 80 percent of the value for that nutrient declared on the label. A product is misbranded if the amount of calories, sugars, total fat, saturated fat, cholesterol, or sodium present is greater than 20 percent in excess of the value for that nutrient declared on the label.

## Percent Daily Value (21 CFR Sec. 101.9(d)(6))

The percent of DV must be declared for each nutrient. The DV is a reference amount based on the daily reference values and reference daily intakes. DVs for fat, saturated fat, cholesterol, and sodium are based on the maximum amounts that people are recommended to consume. The DVs are based on a 2,000 calorie a day diet.

Requiring nutrients to be declared as a percentage of the DV is intended to prevent misinterpretations that arise with quantitative values. For example, 5 g of saturated fat in a food may seem like a small amount if you did not know that 20 g is the daily value for saturated fat.

### Nutrition Panel Footnote (21 CFR 101.9(d)(9))

The percentage of DV listing carries a footnote saying that the percentages are based on a 2,000-calorie diet. Some nutrition labels—at least those on larger packages—have additional footnotes.

## Format Modifications

Deviations from the format discussed above are allowed in a few instances. The tabular and linear formats may be used on packages with less than 40 square inches available for labeling and insufficient space for the full vertical format with more than 40 square inches available for labeling. The linear format may only be used if the tabular format will not fit on the label. These packages may also abbreviate the names of the dietary components, omit footnotes (except for the statement that "% Daily Values are based on a 2,000-calorie diet"), and place the nutrition information on other panels readily seen by the consumer.

A simplified format is allowed on foods containing insignificant amounts of seven or more of the mandatory nutrients and total calories. *Insignificant* means that a declaration of zero could be made in nutrition labeling, or for total carbohydrate, dietary fiber, and protein, the declaration states "less than 1 g".

With this format, total calories, total fat, total carbohydrate, protein, sodium, and any other mandatory nutrients that are present in significant amounts must be listed. The other mandatory nutrients that are not present in significant amounts may be listed after the statement "Not a significant source of. . . ."

The aggregate format may be used when more than one type of packaged food is contained in one package, such as a multipack or variety pack of cereal or snack foods. It may also be used when a single package is to be used for more than one variation of food, such as round ice cream containers.

A dual declaration may be made in cases where a food is commonly combined with another food before being consumed, such as cereal and milk, or when the food will be consumed in a form different from that in the package, such as a cake mix.

## Nutrition Content Claims

Have you ever wondered what the words *light*, *reduced*, and *low* really mean, or if they mean anything at all? According to the FDA, these words and others used in nutrient content claims have very specific meanings, and foods with labels that use them incorrectly are considered misbranded.

The core terms that can be used for nutrient content claims are explained below. Alternate spelling of the terms and their synonyms are allowed as long as the alternatives are not misleading.

### Free

Acceptable synonyms of *free* are *dietarily insignificant source of*, *negligible source of*, *no*, *trivial source of*, *without*, and *zero*.

To be labeled as *free* of a nutrient, the product must contain no amount of, or only a trivial, or physically inconsequential amount of the nutrient. The nutrients that a product can be labeled as *free* of, along with the nutrient levels that are considered physically inconsequential, are listed below (21 *CFR* Sec. 101.60, 101.61, 101.62).

- Calorie free—less than 5 calories per serving (21 *CFR* 101.60(b))
- Sugar free—less than 0.5 grams of sugar per serving (21 *CFR* 101.60(c))
- Sodium free—less than 5 milligrams per serving (21 *CFR* 101.61)
- Fat free—less than 0.5 grams per serving (21 *CFR* 101.62(b))
- Saturated fat free—Less than 0.5 g saturated fat and less than 0.5 g *trans* fatty acids per reference amount and per labeled serving (or for meals and main dishes, less than 0.5 g saturated fat and less than 0.5 g *trans* fatty acids per labeled serving) (21 *CFR* 101.62(c))
- Cholesterol free—less than 2 milligrams per serving (21 *CFR* 101.62(d))

Foods that naturally don't contain a certain nutrient must be labeled to indicate that all foods of that type meet the claim. For example, a fat-free claim on applesauce would have to read "applesauce, a fat-free food," or "applesauce, a naturally fat-free food."

## Low

Synonyms for *low* include such terms as *little*, *few*, *low source of*, and *contains a small amount of*.

A food meets the definition for "low" if a person can eat the food frequently without exceeding the DV for the nutrient. The following specific values are required (21 *CFR* Sec. 101.60, 101.61, 101.62).

- Low-calorie—40 calories or less per serving
- Low-sodium—140 milligrams or less per serving
- Very low sodium—35 milligrams or less per serving
- Low-fat—3 grams or less per serving
- Low-saturated fat—1 gram or less per serving
- Low cholesterol—20 milligrams or less and 2 grams or less of saturated fat per serving

## Lean and Extra Lean

*Lean* and *extra lean* are terms that can be used to describe the fat content of meat, poultry, seafood, and game meats (21 *CFR* Sec. 101.62).

- Lean—less than 10 grams of fat, 4.5 grams or less saturated fat, and less than 95 milligrams cholesterol per serving and per 100 grams
- Extra lean—less than 5 grams of fat, less than 2 grams of saturated fat, and less than 95 mg cholesterol per serving and per 100 grams

## High

Foods that are high in a certain nutrient must contain 20 percent or more of the DV for that nutrient (21 *CFR* Sec. 101.54). Synonyms of *high* include *rich in* and *excellent source of*.

### Good Source

The terms *contains* and *provides* may also be used. To be considered a "good source" of a nutrient, the food must provide 10–19 percent of the DV for the nutrient (21 *CFR* Sec. 101.54).

The next set of claims are called comparison claims or relative claims. These claims compare the food in question with a reference or similar food. Relative claims must include the percentage of difference and the identity of the reference food beside the most prominent claim. Each claim has rules about what can be used as a comparison food (21 *CFR* 101.13).

## Reduced

The reference food used for a reduced claim must be a similar food that is from either the same manufacturer or a different manufacturer. For example, reduced-fat potato chips should be compared with regular potato chips.

A food labeled as *reduced* must contain at least 25 percent less of the nutrient or calories than the reference product. For example, let's say that a serving of potato chips has 12 grams of fat. For modified potato chips to be labeled as *reduced fat*, they must contain 9 grams of fat or less. A reduced claim can not be made if the reference food meets the requirement for a "low" claim for the nutrient in question.

The reference food for a "less" claim may be a similar food or a dissimilar food within a product category. For example, reduced-fat potato chips may be compared with regular potato chips. The food must contain 25 percent less of a nutrient or calories than the reference food.

## More

The reference food for a "more" claim may be a similar food or a dissimilar food within a product category. For foods that are altered, the terms *fortified*, *enriched*, and *added* may also be used. To use this term, the food must contain a nutrient that is at least 10 percent of the DV more than the reference product (21 *CFR* 101.54).

### Light or Lite

Foods that have "light" claims must be nutritionally altered. The reference food for a "light" claim must be a similar food that is not the manufacturers' own product. In addition, the nutrition value for the reference food should come from a value in a valid database, an average value from the top three regional or national brands, a market basket norm, or a market leader (21 *CFR* 101.13, 101.56).

- *Light* may be used for foods that contain 40 calories or less, 3 grams of fat, and 50 percent less sodium than the reference food.
- *Light in calories* may be used when calorie content is 33⅓ percent (one-third) lower than the reference food. The claim may not be made if reference food is low calorie.
- *Light in fat* may be used when fat content must be 50 percent lower than the reference food if the food derives 50 percent or more of calories from fat, or 33⅓ percent lower if the food derives less than 50 percent of calories from fat. If the food contains less than 40 calories or less than 3 grams of fat, the percentage of reduction does not have to be declared. The claim may not be made if reference food is low fat.
- *Light in sodium* may be used if the food contains 50 percent less sodium than the reference food.
- *Lightly salted* may be used if 50 percent less sodium has been added to the reference food.

*Light* may also be used to describe other characteristics of the food as long as qualifying information is included, such as "light and fluffy texture." The term may also be used on foods that have commonly been called light, such as light brown sugar.

## Other Claims

Other claims are also regulated by the FDA. These claims give terms such as *healthy* and *fresh* specific meanings and address special foods such as meals, main dishes, and standardized foods.

### Percent Fat Free

Any product bearing this claim must be a low-fat or a fat-free product. The claim must reflect the amount of fat present in 100 grams of the food.

Thus, if a food contains 2.5 grams of fat per 50 grams, the claim must be "95 percent fat free" (21 *CFR* Sec. 101.62, 101.13).

## Health Claims

A food is considered misbranded if any relationship between a nutrient and disease (a health claim), except for those approved by the FDA, is stated on the label. This was not always the case. Before the NLEA, *all* references to a disease on the label were prohibited. If a relationship was mentioned, the product would cease to be a food and would be regulated as a drug, meaning much more strenuous regulation.

The current regulations, found in 21 *CFR* Sec. 101.14, define a health claim as any claim made on the label or in labeling of a food that characterizes the relationship of any substance to a disease or health-related condition. This includes

- third party references,
- written statements (e.g., a brand name including a term such as "heart"),
- symbols (e.g., a heart symbol), and
- and vignettes.

Health claims can be stated outright or simply implied by the labeling. An example of an allowed health claim is the statement, "While many factors affect heart disease, diets low in saturated fat and cholesterol may reduce the risk of this disease."

### Approved Health Claims

To educate the consumer about recognized diet–disease relationships, the FDA now allows a limited number of health claims on food labels. The nutrient/disease relationships allowed to be mentioned as health claims are

- calcium and osteoporosis (21 *CFR* Sec. 101.72),
- fat and cancer (21 *CFR* Sec. 101.73),
- saturated fat and cholesterol and coronary heart disease (CHD) (21 *CFR* Sec. 101.75),
- fiber-containing grain products, fruits, and vegetables and cancer (21 *CFR* Sec. 101.76),

- fruits, vegetables, and grain products that contain fiber and risk of CHD (21 *CFR* Sec. 101.77),
- sodium and hypertension (high blood pressure) (21 *CFR* Sec. 101.74),
- fruits and vegetables and cancer (21 *CFR* Sec. 101.78),
- folic acid and neural tube defects (21 *CFR* Sec. 101.79),
- sugar alcohols and dental caries (21 *CFR* Sec. 101.80), and
- soluble fiber from whole oats and coronary heart disease (21 *CFR* Sec. 101.81).

For more information about each of these claims use the reference following each claim to look up details about the claim in the *Code of Federal Regulations*.

## Use of Health Claims

The approved health claims may be used only if the following conditions are met:

- The use of the nutrient/disease relationship statement is consistent with the conclusions described in the regulation allowing that statement.
- The claim is limited to describing the value that ingestion (or reduced ingestion) of the substance, as part of a total dietary pattern, may have on a particular disease or health-related condition.
- The claim must be complete, truthful, and not misleading. Where factors other than dietary intake of the substance affect the relationship between the substance and the disease or health-related condition, such factors may be required to be addressed in the claim.
- All information required to be included in the claim must appear in one place without other intervening material, except that the PDP of the label or labeling may bear the reference statement, "See (resource) for information about the relationship between (food item) and (health issue)." In this case the entire claim must appear elsewhere on the label. If graphic material is used, such as a picture of a heart, the above statement or the health claim must appear near the graphic material.
- The claim must enable the public to comprehend the information provided and to understand the relative significance of such information in the context of a total daily diet.

- The food must contain the appropriate amount of the nutrient in question. These amounts are discussed above and can be found in the regulations.
- In addition, foods that use these health claims must not contain disqualifying nutrient levels, which are levels of total fat, saturated fat, cholesterol, and sodium that are above the levels set by the FDA.

There are exceptions to the above levels. As scientific evidence supports more relationships, the FDA will approve other health claims that may be used. The requirements for new health claims are also listed in 21 *CFR* 101.14.

## USDA-Regulated Products

The FSIS, part of the USDA, is responsible for regulating meat and poultry products, including product labeling. FSIS requirements for food labels are similar to FDA requirements in many ways. Like the FDA, FSIS requires that the following elements be present on food labels:

- the name of the product (based on a standard of identity common name, or truthful description of the product);
- a list of ingredients in descending order of presence by weight;
- the name and place of business of the manufacturer, packer, or distributor for whom the product is prepared; and
- a net quantity of contents statement.

Additionally, FSIS requires that an official inspection legend and an official establishment number be present on labels.

FSIS has also defined many terms that processors are allowed to use on their labels. In order to help the consumer understand those terms, FSIS has published *Meat and Poultry Labeling Terms*. As with FDA-regulated foods, it is very important for all the ingredients in the foods to be listed on the ingredient statement. Persons with food allergies rely on these statements when making food choices.

### Nutrition Labeling

FSIS requires that meat and poultry products have nutrition labeling similar to that of FDA-regulated foods. The required labeling includes the

Nutrition Facts box used on FDA-regulated foods. These regulations are located in 9 *CFR* Sec. 317.300–500. FSIS has also adopted FDA's restrictions on nutrient content claims and health claims.

## Mandatory Safe-Handling Statements

Safe-handling statements are required for many FSIS-regulated foods. Regulations requiring these statements are located in 9 *CFR* Sec. 317.2(k)(l) and 381.125. These products must have prominently displayed on the PDP of the label the statement: "Keep Refrigerated," "Keep Frozen," "Perishable Keep Under Refrigeration," or such a similar statement as the FSIS may approve in specific cases.

Products that contain meat or poultry ingredients and that are intended for "household consumers, hotels, restaurants, or similar institutions" must contain the following disclaimer:

> This product was prepared from inspected and passed meat and/or poultry. Some food products may contain bacteria that could cause illness if the product is mishandled or cooked improperly. For your protection, follow these safe handling instructions.
>
> The following statement must appear:

1. Keep refrigerated or frozen. Thaw in refrigerator or microwave. (Any portion of this statement that is in conflict with the product's specific handling instructions, may be omitted, for example, instructions to cook without thawing.) (A graphic illustration of a refrigerator shall be displayed next to the statement.)
2. Keep raw meat and poultry separate from other foods. Wash working surfaces (including cutting boards), utensils, and hands after touching raw meat or poultry. (A graphic illustration of soapy hands under a faucet shall be displayed next to the statement.)
3. Cook thoroughly. (A graphic illustration of a skillet shall be displayed next to the statement.)
4. Keep hot foods hot. Refrigerate leftovers immediately or discard. (A graphic illustration of a thermometer shall be displayed next to the statement.)

## Summary

For the most part, labeling regulations are similar for FDA- and USDA-regulated products, with the exception of those discussed previously. The regulations discussed in this chapter were current when this book was published, but regulations do change frequently. Therefore, Internet resources should be used to determine the most current information.

The *Food Labeling Guide*, *Food Labeling Questions and Answers*, *Food Labeling Questions and Answers Volume II*, and the Small Business Food Labeling Exemption are available in the "food" section of the FDA's Internet website. These are invaluable resources and were developed by the government specifically to address, in detail, the NLEA requirements.[1]

## References

1. FDA. 1999. A Food Labeling Guide. U.S. Food and Drug Administration, Center for Food Safety & Applied Nutrition. Online http://www.cfsan.fda.gov/~dms/flg-toc.html.
2. 21 *CFR* Part 101 Food Labeling. Online http://www.access.gpo.gov/nara/cfr/waisidx_01/21cfr101_01.html (September 2004).
3. FDA. 1999. A Food Labeling Guide. Chapter II, Name of Food. FDA Center for Food Safety and Applied Nutrition. Online http://www.cfsan.fda.gov/~dms/flg-2.html (September 2004).
4. FDA. 1999. A Food Labeling Guide. Chapter III, Net Quantity of Contents. FDA Center for Food Safety and Applied Nutrition. Online http://www.cfsan.fda.gov/~dms/flg-3.html(September 2004).
5. FDA. 2001. Compliance Policy Guide Sec. 555.250 Statement of Policy for Labeling and Preventing Cross-Contact of Common Food Allergens. Online http://www.fda.gov/ora/compliance_ref/cpg/cpgfod/cpg555-250.htm (September 2004).
6. FDA. 2001. Questions and Answers on Allergen Guides. Online http://www.cfsan.fda.gov/~dms/alrgpgtp.html (September 2004).
7. FSIS. Labeling and Consumer Protection Allergens—Voluntary Labeling Statements. Online http://www.fsis.usda.gov/Frame/FrameRedirect.asp?main=/oppde/larc/ingredients/allergens.htm (September 2004).
8. FDA. 2004. Prior Notice of Imported Food Questions and Answers (Edition 2), Guidance for Industry. Online http://www.cfsan.fda.gov/~pn/pnqagui2.html#intro (September 2004).

9. FSIS. 2000. Mandatory Country of Origin Labeling of Imported Fresh Muscle Cuts of Beef and Lamb. Online http://www.fsis.usda.gov/Frame/FrameRedirect.asp?main=http://www.fsis.usda.gov/oa/Congress/cool.htm (September 2004).
10. FDA. 2003. Industry Guidance Food Labeling: Trans Fatty Acids in Nutrition Labeling, Nutrient Content Claims, and Health Claims. FDA Small Entity Compliance Guide. Online http://www.cfsan.fda.gov/~dms/transgui.html (September 2004).

CHAPTER 6

# Environmental Regulations and the Food Industry

Theodore A. (Ted) Feitshans, North Carolina State University

Any business in the food industry must follow all of the environmental laws that any other business or individual must follow. For the purposes of this chapter, these laws are organized into discharges to surface and groundwaters; solid waste; hazardous waste; use of water; regulation of water sources; discharges to air; chemical use, storage, release, transport, and reporting thereof; and siting and operation of facilities. While federal laws are relatively uniform across geographic regions, state and local laws vary dramatically. Federal laws will be covered by name, whereas, for the most part, state and local laws will be covered generically using relevant examples. This chapter will conclude with a discussion of environmental risk management as a tool to reduce risks to the environment and the likelihood of regulatory violations.

## Discharges to Surface and Groundwaters

The Clean Water Act (CWA), also called the Federal Water Pollution Control Act (33 *U.S.C.* §§1251 to 1387 (2003)), governs all direct discharges to surface waters. A direct discharge is broadly defined to include discharges to ditches that eventually make their way to surface waters. Such discharges may include water used for washing, water extruded from food products during processing, and any other liquid wastes. There is in the CWA an exemption for agricultural storm water discharges; however, that exemption will almost never be applicable to the food industry. The agricultural storm water exemption is not defined in the CWA; however, courts have restricted its application to primary producers of crops and livestock who are engaged in agricultural activities that involve extensive acreages and diffuse runoffs.

When Congress enacted the CWA in 1972, it established the National Pollutant Discharge Elimination System (NPDES) permit program, applicable to all point source dischargers. All point source dischargers have an affirmative duty to obtain a permit prior to making any discharges.

Under the structure of the CWA, states are encouraged to develop their own permit programs under delegation from the Environmental Protection Agency (EPA). The basic requirement for delegation is that no delegated state program can be less stringent than that required by the CWA. States are allowed to develop more stringent programs than those required by the CWA, and many states have developed programs that are much more stringent. All states, except Alaska, Idaho, Massachusetts, New Hampshire, and New Mexico, have approved NPDES permit programs (EPA 2004, http://cfpub.epa.gov/npdes/statestats.cfm). Somewhat fewer states have approved pretreatment programs, and only Oklahoma, South Dakota, Texas, Utah, and Wisconsin have approved biosolids programs.

There are two basic types of permits, general and individual. General permits are developed to cover a type of activity that has a relatively minor impact on water quality. For example, the EPA has issued a general permit covering Alaska seafood processors (EPA NPDES Permit No. AK-G52-0000). The permit lists the wastes that may be discharged, the geographic areas excluded from coverage under the permit, the categories of processors included, record-keeping requirements, and other conditions. As with most general permits, anyone seeking coverage must apply by filing a notice of intent (NOI) with the EPA. Where the general permit is issued by a state under delegated authority from the EPA, the NOI would be filed with the state agency that received the delegation from the EPA. Each general permit will specify the information to be included in the NOI. This information will usually include existing NPDES permit coverage for the facility, complete ownership and contact information, operator information (if different from the owner information), information about the facility, information about the water that will receive the discharge, information about the discharges to be made, and appropriate signatures of those with responsibility for the facility. Some general permits may require that a public hearing be held prior to issuance of a notice of inclusion. When the EPA or the appropriate state agency issues a notice of inclusion, the facility may generally begin operating.

Many, if not most, food production facilities that discharge will be required to file for an individual permit. The information requirements for individual permit applications are more detailed than for applications under general permits. Necessary forms are available on the EPA website (EPA 2004, http://cfpub.epa.gov/npdes/). Most applicants will require expert assistance to help them through the process.

The difficulty of obtaining a permit will depend in part on whether the water to which discharges are to be made is impaired. Waters that are impaired are polluted with one or more pollutants to the extent that the water no longer supports its designated use. For example, a body of water might be classified as supporting fishing for warm-water species. If the body of water is impaired, it will generally have a total maximum daily load (TMDL) established for each pollutant that is causing impairment. A TMDL is the maximum amount of pollutants from all sources that a body of water is capable of receiving without being impaired.

Discharging to a municipal sewerage system does not relieve a food business of its permit obligations. Municipal sewage systems are not designed to handle many materials that food-processing plants may discharge. Toxic materials in the waste stream may kill microorganisms that are essential to the proper operation of municipal treatment systems with the result that sewage may be discharged without being fully treated. Contamination of sludge may limit disposal options and increase costs. While food-processing facilities may not be the most important source of toxic substances, food-processing plants are major contributors of grease and biological oxygen demand (BOD). Grease has the unfortunate tendency to clog pipes and produce spills of raw sewage. BOD increases the load on municipal sewage plants, making it more difficult and expensive to process waste. For these reasons most food-processing facilities will be required to pretreat waste before it is discharged to a municipal sewer system. Information about the National Pretreatment Program may be found on the EPA website (EPA 2004, http://cfpub.epa.gov/npdes/home.cfm? program_id=6). Operators and owners of food-processing facilities should contact the operator of their municipal sewage system before making discharges to it. Many system operators have requirements in addition to those mandated by the EPA.

In addition to coverage under an NPDES permit for discharges from the food-processing operation, a facility may also have to be permitted under the NPDES program for storm water runoff from the facility. Storm water

**Table 6.1.** Standard industrial classification (SIC) code

| *SIC Code* | *Description* |
|---|---|
| 405 | Dairy products processing |
| 406 | Grain mills |
| 407 | Canned and preserved fruit and veg. processing |
| 408 | Canned and preserved seafood processing |
| 409 | Beet, crystalline, and liquid cane sugar refining |
| 20 | Food and kindred product |
| 4222 | Refrigerated storage |

runoff from parking lots, roofs, and other hard surfaces must, under certain conditions, undergo treatment before it is discharged to surface water (EPA 2004, http://cfpub.epa.gov/npdes/home.cfm?program_id=6). The requirement applies whether the discharges are to a municipal separate storm sewer system (MS4) or directly to a surface water. Coverage is determined by standard industrial classification (SIC) code. Examples of food businesses that are covered can be found in Table 6.1.

Construction of these facilities may also require an NPDES storm water permit. Greater detail is available from the EPA website (EPA 2004, http://cfpub.epa.gov/npdes/home.cfm?program_id=6).

## Solid Waste

Most food-processing and related businesses will generate considerable quantities of solid waste, including nonedible portions of plants, spoiled products, sludge, and other residues. At the federal level, disposal of solid waste is governed by the Solid Waste Disposal Act (42 *U.S.C.* §§ 6901–6992k (2003)). Solid waste disposal is also governed by a host of state and local regulations; however, the Solid Waste Disposal Act sets the minimum requirements for solid waste disposal.

The first consideration when evaluating solid waste is to determine that it is not a hazardous waste covered under the Resource Conservation and Recovery Act (RCRA) (subchapter III of the Solid Waste Disposal Act [42 *U.S.C.* §§ 6921–6939e]). Many substances that are RCRA hazardous wastes have been listed by the EPA; however, a waste need not be listed to be hazardous under RCRA. Any waste that exhibits toxicity, flammability, corrosiveness, or reactivity may be hazardous wastes that require special handling under RCRA. A waste that is toxic is one that is likely to poison human, animals, fish, or vegetation. A waste that is flammable is one that is likely to catch fire. This is of special concern in the food industry because some wastes may catch fire spontaneously if not properly handled. Wastes that are corrosive tend to eat away their containers and can cause severe injury to those handling or otherwise exposed to the waste. Wastes that are reactive are unstable and may explode or spontaneously undergo other reactions. Handling of hazardous wastes will be discussed in the next section. Generators of small quantities of hazardous wastes may, under certain circumstances, be exempt from the special handling requirements of RCRA; however, many states, localities, and landfill operators have more stringent rules, thus it is incumbent upon the waste generator to know those rules and proceed accordingly.

Most nonhazardous solid waste in the United States is handled by disposal in a licensed landfill. Recognizing that land for landfills is limited, Congress directed the EPA to develop programs to protect the resource. These programs include source reduction, recycling, and diversion of certain wastes from landfill facilities. Source reduction involves evaluation of the production process to reduce the amount of waste generated. Source reduction programs have an added advantage to processors: less waste to be disposed means less cost of disposal and less raw material wasted. Sometimes source reduction may be achieved by finding new uses for material that was previously a waste product. For example, orange juice producers discovered, many years ago, that the pulp left after extraction of the juice was an excellent cattle feed. Thus a waste product that cost money for disposal was converted into a product that generated additional revenue. Recycling involves recovering useable product from waste. For example, aluminum beverage containers can be collected and melted to produce aluminum at a cost far lower than that associated with the initial production of aluminum. As these examples demonstrate, there is not a firm line between source reduction and recycling.

Some wastes can be diverted from the waste stream for special treatment. These are organic wastes that may be composted. Compost has many uses, including some in other pollution-prevention programs (EPA 2004, http://www.epa.gov/epaoswer/non-hw/compost/index.htm).

## Hazardous Waste

Generators of hazardous waste must apply to the EPA for an EPA identification number (EPA 2004, http://www.epa.gov/rcraonline/index.htm). The application must identify the site, the site land type, the North American Industrial Classification System Code(s) for the activity carried out on the site, the contact person, and the legal owner of the site. The hazardous waste activities that occur on the site must be identified. Generation of hazardous waste must be classified by the quantity produced annually. Those who transport, treat, and dispose of hazardous waste are also regulated. While food businesses are less likely than most to be generators of hazardous waste, the activities conducted must be carefully evaluated to determine whether any are regulated under the RCRA. For example, fluorescent lightbulbs contain mercury. Large sites that dispose of significant quantities (more than 100 kilograms annually) must generally treat spent bulbs as hazardous waste. State and local regulations may impose regulations that are more stringent than the federal requirements. In general, regulated quantities of hazardous waste must be disposed of in a licensed hazardous waste landfill or by other approved means. A manifest must follow the waste from the site where it is generated to the disposal site. A regulatory agency should be able to find a complete paper trail for the waste. Failure to observe these regulations is likely to result in the imposition of civil penalties. Willful violations may result in the imposition of criminal penalties.

## Use of Water

Water use is generally regulated under state and local law rather than under federal law. Sources of water are either surface water or groundwater. The source of the water will affect the regulatory scheme. Most states have elaborate systems for dividing surface water. In eastern states this system is generally one of riparian rights. Only riparian owners have the right to use surface water. A riparian owner is one whose property touches the surface body of water in question. An owner whose property is near the body of water, but does not touch it, is not a riparian owner and has no right of use. Under early English law the ability of a commercial enterprise to use surface water was sharply limited. Commercial enterprises were not allowed to diminish either the quality or the quantity of the water. Quality (that is discharge of pollutants) is now regulated under the federal CWA. Quantity, except where water is supplied by federal water projects, is still largely regulated at the state level.

In eastern states, the English rule, applicable to commercial enterprises, has generally been modified to a rule of reasonable use. That means that the use of surface water is limited to quantities that are reasonable under the circumstances. It is often very difficult for a water user to determine in advance just what quantity would be a reasonable quantity. The picture of water use in eastern states is further complicated by the development of regulatory schemes in some states that are designed to divide water supplies among competing users. Further complicating the picture of eastern water use is the existence of deeds of easement that purport to extinguish the right of one riparian owner to use water in favor of another riparian user on the same body of water.

Artificial bodies of water further complicate the picture of eastern water use. Artificial bodies of water are impoundments, lakes, ponds, canals, and other bodies of water created by human intervention. Owners of properties that touch older artificial bodies of water are often treated as riparian owners with full rights to use the water. Where the artificial bodies of water are relatively new, the abutting landowners may not be riparian owners and may have no right of water use. Often, use of water from artificial reservoirs is governed by deed restrictions or federal, state, or local regulations applicable only to the specific reservoir in question.

Groundwater in eastern states is generally treated in the same manner as surface water if the water is found in well-defined underground channels. Most groundwater, called *percolating groundwater*, is found in the pores in the soil or rock found below the water table. It is not found in defined channels. Under English common law, it was treated as the property of the landowner under a rule of absolute ownership. However, most states have modified this rule to one of reasonable use in recognition of the fact that what one landowner does with the groundwater under her property will have an impact on the groundwater under neighboring properties. Some states have developed regulatory systems (including licensing requirements) controlling the use of groundwater, particularly for large commercial users.

Regulation of water quantity in western states is radically different from regulation of water quantity in eastern states. Most western states use a system of appropriative rights under which the first user in time is first in right. If there is no water left after the first user uses his share, then the users with inferior rights are simply out of luck.

The right to use water may also be limited by international treaty, interstate compacts, treaties with Indian Nations, and the Endangered Species Act (ESA). Where a water source is a habitat for a species that is listed as endangered or threatened under the ESA, water users may be required to limit their water use so that sufficient water remains in the water source to provide adequate habitat for the listed species. Since water is of critical importance to many food industry businesses, it is incumbent upon anyone developing such a business to fully explore the myriad of legal restrictions on their proposed water source.

## Regulation of Water Sources

Surface water sources for public water supply systems, as defined under the federal Safe Drinking Water Act (SDWA), are highly regulated. The SDWA requires that the local zoning authority establish land use restrictions in the watershed that supports the surface water source. These specially restricted areas are called water supply watershed protection programs. Restrictions in water supply watershed protection areas are often complex and vary greatly from one water supply watershed protection area to another. Regulations in water supply watershed protection areas may include density restrictions and other restrictions on construction. Land uses that are likely to contaminate a surface water source may be either banned or restricted. The percentage of hard surface, including parking lots, roofs, and other surfaces that result in increased runoff, may be restricted.

The SDWA encouraged, but did not require, that states develop wellhead protection programs to protect groundwater sources supplying public water supply systems. Like water supply watershed protection programs designed to protect surface water supplies, wellhead protection programs are designed to restrict land uses on the land above the aquifer. Some land uses may result in the release of pollutants that eventually are carried into the groundwater. These restrictions limit the uses to which land in the wellhead protection area can be put.

Many states and local governments establish standards for wells; however, a surprisingly large number of jurisdictions provide for no regulation of wells whatsoever. Food businesses that expect to use wells as their source of water must, of course, comply with state and local well regulations. Those businesses locating where well regulations do not exist, may, however, have the more difficult task. Where the water is ultimately incorporated into food products, contaminants in the water may result in the food product being adulterated under the federal Food, Drug, and Cosmetics Act. It is therefore incumbent upon the food producer to contract with reliable well drillers and require, by contract, that an adequate well, adequately sealed against surface contamination, be provided. Such contracts should require adequate insurance coverage on the part of the well driller and should also include clauses providing for indemnification of the food producer's expenses should the well be inadequate for the needs of the food producer. Food processors should also periodically test the water that they are using for major contaminates that might cause a food product to be deemed adulterated.

## Discharges to Air

Food-processing and related businesses may make discharges to air. These discharges are highly regulated at both the federal and state level. The federal Clean Air Act is the basic legislation under which responsibilities for the protection of air quality are divided between the EPA and the states. The regulatory scheme under the Clean Air Act is one of the most complex in existence. Disagreements over various aspects of the Clean Air Act continue to engender a massive amount of litigation. Compliance is difficult and costly for those businesses that are covered. What follows is a very brief explanation of the regulatory program.

Criteria pollutants under the Clean Air Act are carbon monoxide, lead, nitrogen oxides, ozone, sulfur dioxide, and particulates, both fine and course. The EPA has established national ambient air quality standards (NAAQS) for each criteria pollutant. Primary standards are designed to protect human health, whereas secondary standards are designed to protect the public welfare. Not all criteria pollutants have established secondary standards.

For each criteria pollutant states are expected to develop state implementation plans (SIPs). SIPs are incredibly complex regulatory arrangements designed to reduce specific criteria pollutants. A wide range of control measures may be considered by a state when developing a SIP. These measures include emissions controls on polluters but may also extend to such measures as controls on development; carpooling arrangements; restricted, high-occupancy lanes; and other indirect measures designed to reduce mobile source pollution.

An area that is not meeting a NAAQS for a criteria pollutant is called a nonattainment area. The EPA may force states to impose additional restrictions on such areas. These restrictions may be broad ranging, including development controls. Such limitations may extend to individual consumers through restrictions on backyard grilling and the use of wood-burning stoves.

The Clean Air Act does not address all air quality issues. The use of chlorofluorocarbons (CFCs) in refrigeration equipment is addressed under the Montreal Protocol. Under the provisions of this convention, refrigeration equipment has largely been converted to refrigerants with less potential for damaging the ozone layer of the atmosphere. Technicians who install and repair refrigeration and air-conditioning equipment must have proper training and licensing.

Indoor air pollutants may also not be addressed under the Clean Air Act. Radon, a naturally occurring, radioactive gas, is addressed under the 1986 Radon Gas and Indoor Air Quality Act. Other indoor air pollutants, found in the workplace, may be addressed under the federal Occupational Safety and Health Act, covered in chapter. 7

The Clean Air Act also does not cover greenhouse gases such as carbon dioxide and methane, which contribute to global warming. The United States has not ratified, and has officially expressed its opposition to, the Kyoto Protocol that was negotiated to address this issue. The Kyoto Protocol has not been ratified by sufficient industrial nations to enter into force. Although it has not become law, and would not be law in the United States, it is, nonetheless, an issue for food businesses. The current Administration has developed voluntary programs to encourage reduced emissions, and some potential customers may impose requirements on their suppliers. Customers located in the European Union or Japan are particularly likely to impose restrictions on suppliers.

## Chemical Use, Storage, Release, Transport and Reporting There Of

### Toxic Substance Control Act

A premanufacture notification (PMN) must be submitted to the EPA for each new chemical substance that a company intends to manufacture or import (15 *U.S.C.* §2604 (2003)). This notice must be submitted at least 90 days before importation or manufacture. A new chemical substance is one that has not been listed by EPA under section 2607(b). Since some submissions are held confidential by the EPA to protect company trade secrets, a potential applicant may request a Toxic Substance Control Act (TSCA) inventory search if the potential applicant can prove a bona fide intent to manufacture or import the chemical (40 *C.F.R.* §720.25 (2003)).

There are important exceptions to the requirement that a PMN be submitted. Mixtures that are mere combinations of existing chemicals need not be submitted under a PMN. Tobacco and tobacco products are exempt. Pesticides regulated under the federal Insecticide, Fungicide, and Rodenticide Act are exempt. Any food, food additive, drug, cosmetic, or device regulated under the federal Food, Drug, and Cosmetic Act is exempt.

Manufacture for the purposes of TSCA includes imports; however, articles (those things that are already manufactured) do not require a PMN even though a new chemical may be contained within an article. Submitters must submit all of the data that they have about a new chemical but generally need not develop any new data nor conduct any toxicity studies. For employees in large organizations who are preparing PMNs it may be necessary to submit proof that a diligent search was made of all of the organization's records that might relate to the new chemical.

## Community Right-to-Know

The 1984 disaster at the Union Carbide pesticide plant at Bhopal, India, motivated Congress to enact the Emergency Planning and Community Right-to-Know Act (EPCRA) of 1986 as Title III of the Superfund Amendments and Reauthorization Act of 1986 (SARA Title III). That disaster killed at least 20,000 people and sickened another 120,000 (The International Campaign for Justice in Bhopal 2004, http://www.bhopal.net/index.php). Congress discovered, as a result of its investigation, that there was no reason that the disaster could not have happened in the United States and that local communities were completely unprepared to mount an adequate response.

There are three components to the EPCRA: emergency notification and planning, release reporting, and general provisions applicable to information collecting and reporting. Local emergency responders and their counterparts in industrial facilities within their jurisdiction are required to plan for emergencies. Owners or operators of facilities are required to immediately report all unplanned releases of hazardous chemicals. Under the second component of EPCRA all industrial facilities are required to annually report all releases of hazardous chemicals. The EPA publishes these inventories annually. Because the definitions in the EPCRA are quite broad, the coverage under EPCRA extends to facilities (such as community swimming pools) that one does not ordinarily think of as industrial. Coverage under EPCRA is sufficiently broad to reach most food businesses. Reporting requirements under EPCRA are in addition to and independent of reporting requirements under other federal and state environmental laws.

EPCRA envisions a division of responsibility between federal, state, and local authorities, with most of the authority and responsibility residing at the state and local levels. EPA's responsibilities are confined primarily to three areas: The first is reviewing claims for trade secrets for information not included in required reports; the second is preparing the annual Toxic Chemical Release Inventory report; and the third is the EPA's National Response Center receiving some reports of accidental releases of hazardous chemicals.

The EPA also developed the list of covered chemicals and the threshold levels for coverage. Any facility that contains a covered chemical at a threshold level or higher must comply with EPCRA. Exempt facilities include vessels, motor vehicles, rolling stock, and aircraft. These are all covered under other laws; however, releases from these exempt facilities must nonetheless be reported under EPCRA.

Planning for emergencies is primarily a state and local government responsibility. It is the responsibility of each state to enact the necessary state laws to ensure that each local emergency responder has the information, training, and adequate advance planning to respond to any emergency release within their jurisdiction. States were required to establish state emergency response commissions, emergency planning districts, and local emergency planning committees (42 *U.S.C.* § 11001 (2003)). This requirement has proven to be expensive and problematical for state and local governments. Many rural jurisdictions are poorly funded and rely upon volunteer services that lack the wherewithal to fully comply with the requirements of EPCRA. This situation also places a burden on covered facilities to correctly identify the required reports and the entities to whom those reports must be provided.

Section 11002 requires that the owner or operator of a facility where an extremely hazardous substance becomes present at a threshold level (or the existing list of such substances is revised) must report this to the state emergency response commission and local emergency planning committee. It is then the responsibility of the state emergency response commission to report this information to the EPA. The owner or operator of a covered facility must also notify the local emergency planning committee of the identity of the facility emergency planning coordinator who will participate in the emergency planning process (42 *U.S.C.* § 11003(d) (2003)). If there is no local emergency planning committee (or it is inactive), EPCRA requires that the owner or operator of the facility report this information to the governor of the state where the facility is located. A new report must be made each time that the identity of the coordinator (or the coordinator's contact information) changes.

At the heart of EPCRA is emergency release notification (42 *U.S.C.* § 11004 (2003)). Any time that there is an accidental release of a covered chemical, this provision is triggered. Notification is given to the National Response Center, the state emergency response commission, the local emergency planning committee, and to those people affected. Affected people may be notified by radio, telephone, or in person.

Section 11021 requires that the owner or operator of a facility maintain a material safety data sheet (MSDS) for each chemical for which a MSDS is required under the Occupational Safety and Health Act of 1970 (OSHA) (42 *U.S.C.* § 11021 (2003)). The EPA may modify the OSHA requirements regarding thresholds. Of relevance to agriculture and the food business are several exceptions to the MSDS maintenance and reporting requirement.

> (1) Any food, food additive, color additive, drug or cosmetic regulated by the Food and Drug Administration. . . .
>
> (4) Any substance to the extent it is used in a research laboratory . . . under the direct supervision of a technically qualified individual.
>
> (5) Any substance to the extent that it is used in routine agricultural operations or is a fertilizer held for sale by a retailer to the ultimate customer. (42 *U.S.C.* § 11021(e) (2003)).

The owner or operator of a facility covered under EPCRA must submit a MSDS for each covered chemical or a list of those chemicals to the state emergency response commission, the local emergency planning committee, and the fire department with jurisdiction over the facility. When the owner or operator of a covered facility submits a list of chemicals, the local emergency planning committee may request a MSDS for any chemical on the list. Any member of the public may request a copy of the MSDS from the local emergency planning committee. If the committee does not have a copy of the requested MSDS, then the committee must request that the owner or operator of the covered facility provide a copy (42 *U.S.C.* § 11021(c)(2003)).

Owners and operators of covered facilities must prepare an emergency and hazardous chemical inventory for each chemical for which an MSDS is required (42 *U.S.C.* § 11022(2003)). An inventory form containing Tier I information for each such chemical must be submitted to the state emergency response commission, the local emergency planning committee, and the fire department with jurisdiction over the facility. Tier I information includes the following.

> (i) An estimate (in ranges) of the maximum amount of hazardous chemicals in each category present at the facility at any time during the preceding calendar year.
> (ii) An estimate (in ranges) of the average daily amount of hazardous chemicals in each category present at the facility at any time during the preceding calendar year.
> (iii) The general location of hazardous chemicals in each category (42 *U.S.C.* § 11022(d)(1)(B)(2003)).

Tier II information is provided only upon the request of the state emergency response commission, the local emergency planning committee, and the fire department with jurisdiction over the facility 42 *U.S.C.* § 11022(e)(1)(2003). Tier I information includes the following.

> (A) The chemical name or the common name of the chemical as provided in the material safety data sheet.
> (B) An estimate (in ranges) of the maximum amount of the chemical present at the facility at any time during the preceding year.
> (C) An estimate (in ranges) of the average daily amount of the chemical present at the facility at any time during the preceding year.
> (D) A brief description of the manner of storage of the hazardous chemical.
> (E) The location at the facility of the hazardous chemical.
> (F) An indication of whether the owner elects to withhold location information of a specific hazardous chemical from disclosure to the public under section 11044 of this title (42 *U.S.C.* § 11022(d)(2) (2003).

Tier I or II information may be made available to other state and local officials upon request to the state emergency response commission or the local emergency planning committee.

As a matter of policy, Tier II information is generally available to the public upon request; however, the request must be in writing and the requesting member of the public may be required to state the need for the information. Such information may not be available either upon trade secret or homeland security considerations.

Section 11023 requires that the owner or operator of each covered facility prepare a toxic chemical release form at least annually (42 *U.S.C.* § 11023(a)(2003)). Covered facilities include those with 10 or more full-time employees in SIC Codes 20–39 (42 *U.S.C.* § 11023(b)(1)(A)(2003)). The EPA was given authority to add or delete SIC codes as it deemed appropriate, and both the EPA and the governors of the states were given the authority to add additional facilities. Chemicals covered under EPCRA include those specifically listed by Congress, subject to additions and deletions made by the EPA. Any person or the governor of a state may petition the EPA to add or delete a chemical from the list of covered chemicals. EPCRA also established thresholds, subject to revision by the EPA, for reportable quantities of toxic chemicals released. The information required in a toxic chemical release form must

(A) provide for the name and location of, and principal business activity at, the facility;
(B) include an appropriate certification, signed by a senior official with management responsibility for the person or persons completing the report, regarding the accuracy and completeness of the report; and
(C) provide for the submission of each of the following items of information for each listed toxic chemical known to be present at the facility:
   (i) Whether the toxic chemical at the facility is manufactured, processed, or otherwise used, and the general category or categories of use of the chemical.
   (ii) An estimate of the maximum amounts (in ranges) of the toxic chemical present at the facility at any time during the preceding calendar year.
   (iii) For each waste stream, the waste treatment or disposal methods employed, and an estimate of the treatment efficiency typically achieved by such methods for that waste stream.
   (iv) The annual quantity of the toxic chemical entering each environmental medium (42 *U.S.C.* § 11023(g)(2003)).

The toxic chemical inventory release report must be provided to the EPA and the designated state agency for the state where the facility is located.

EPCRA provides that owners and operators of covered facilities may protect trade secrets by omitting information about specific chemicals, where such would normally be required, provided that information about the general category of chemical is provided (42 *U.S.C.* § 11042(2003)).

It is incumbent upon the person claiming a trade secret to provide documentation in support of such a claim. Such documentation includes the following.

(1) Such person has not disclosed the information to any other person, other than a member of a local planning committee, an officer or employee of the United States or a State or local government, an employee of such person, or a person who is bound by a confidentiality agreement, and such person has taken reasonable measures to protect the confidentiality of such information and intends to continue to take such measures.

(2) The information is not required to be disclosed, or otherwise made available, to the public under any other Federal or State law.

(3) Disclosure of the information is likely to cause substantial harm to the competitive position of such person.

(4) The chemical identity is not readily discoverable through reverse engineering (42 *U.S.C.* § 11042(b)(2003)).

Such information may not be withheld from any health professional with a need for the information (42 *U.S.C.* § 11042(e)(2003)). Recognized needs include diagnosis or treatment, medical emergency, and preventive measures (42 *U.S.C.* § 11043(2003)). The person owning the trade secret may demand that the health professional sign a confidentiality agreement, except in the event of a medical emergency. In the event of a medical emergency, the health professional may be required to sign a confidentiality agreement as soon as conditions permit.

## Siting and Operation of Facilities

Zoning and local land use issues must also be considered when siting and operating a food production facility. While these issues often seem peripheral to the food production business, they may determine the success or failure of the entire enterprise.

Most large cities, as well as many small cities and counties, have developed zoning codes. In theory, zoning codes are designed to protect property values by reducing conflicts between land uses. Most, but not all, zoning codes seek to accomplish this goal by grouping compatible land uses together. Zoning codes of larger municipalities may run to hundreds of pages with dozens of zones and subzones. Zoning codes of some small cities and rural counties may be only a few pages long with a half dozen zones. Some zoning codes are very well-written, while others are nightmares of ambiguity.

Failure to abide by zoning regulations may have very serious consequences, including restoration of a property to its original condition. The fact that a governing authority has failed to enforce provisions of its zoning regulation for many years is generally no defense to an action to enforce a zoning regulation. Unlike virtually all other laws, there are generally no statutes of limitation for zoning violations.

Where existing zoning does not meet the needs of a proposed food business, it may be necessary to petition the governing authority for either a zoning variance or a rezoning of the property. This is usually initiated by petitioning the governing authority's zoning board. Zoning boards usually provide the public and, in particular, adjoining landowners, with an opportunity to be heard prior to rezoning or granting a variance to existing zoning. Developing good relationships with neighboring property owners is often critical to the success of this process.

Building codes, fire codes, and other local ordinances must also be reviewed when developing or expanding a food business. Increasingly, zoning ordinances, building codes, and other local ordinances are being used to accomplish environmental goals of local governments. These goals include meeting storm water management requirements (discussed in this chapter), preserving open space, protecting trees and other vegetative features, and reducing noise pollution.

At common law, each landowner has rights to protect the possession of his property by maintaining a trespass action and to protect the use and quiet enjoyment of his property through a nuisance action. These common law torts have increasingly figured into environmental disputes (Richardson and Feitshans, 2000, Nuisance Revisited after *Buchanan* and *Bormann*, *Drake Journal Agricultural Law*, 5, no. 1). Trespass suits have been used for such environmental issues as smoke, odors, and groundwater and surface water contamination. Tort actions based upon a nuisance theory have been similarly used. In both types of actions both damages and injunctive relief are available. Generally, punitive damages are also available. Losing such a lawsuit may be both expensive and crippling to a food business.

## Environmental Risk Management

Environmental laws that govern food processing and other food-related businesses are exceedingly complex. Some of these laws and regulations are ambiguous or even internally inconsistent. Reasonable people often differ as to the requirements and implications of these laws and regulations. Federal, state, and local regulations may have inconsistent or contradictory requirements. Enforcement is often uneven at best. Given such an environment of uncertainty, every business must formulate an environmental risk management strategy as part of its overall compliance plan.

There are five basic steps in any environmental risk management plan. These steps are risk identification; risk evaluation; risk treatment; risk selection and implementation; and risk monitoring. No environmental risk management plan can succeed if it fails to identify the environmental and regulatory risks that it faces. The risks identified should include both physical risks (e.g., the potential for an inadvertent discharge of waste to surface water) and regulatory risks (e.g., the failure to make a required report). An important tool for identifying risks is a good system of incident monitoring and reporting. An incident is something that occurs that could give rise to an event. An event is anything that gives rise to a loss. Typically, before an event occurs, there are warning signs that, if heeded, could result in avoidance of the risk. For example, in an operation that uses a lagoon system for handling liquid waste, there is typically a freeboard requirement for the lagoon. Waste levels exceeding the freeboard requirement are incidents (unless in themselves a regulatory violation.) Waste that actually escapes the lagoon, causing a discharge to surface water (and a violation), is an event. With a good incident-reporting system and the ability to take prompt corrective action, violations and the attendant costs may be avoided.

Risks are evaluated primarily on two criteria: first, the probability of an event, and second the severity of an event once it occurs. The probability of an event occurring is based upon frequency. The severity of the event is usually defined in monetary terms, but should also include the risk of criminal prosecution, both of the business, if organized as a separate legal entity such as a corporation, and employees and officers of the business. Analysis of severity should also include an analysis of the impact that the event is likely to have on the good will of the business, including customers, potential customers, and the general public. The risks of greatest concern are those with the greatest severity, and among risks in the same severity class, the highest probability of occurrence.

Once risks have been identified and evaluated, risk treatment options should be identified. There are usually many such options among which to choose. Each realistic option should be evaluated on the basis of potential effectiveness at eliminating or mitigating the risk, cost, negative consequences, and other relevant factors. Once risk treatments have been identified and analyzed, the best one for a particular business should be selected and implemented.

Environmental risks are typically identified and evaluated through an environmental audit process. Environmental audits, while often essential, carry considerable risks. While some states provide immunity for environmental audits, and some states go further to encourage environmental compliance by providing immunity when violations are voluntarily reported, there are often contradictory rules among states and with the federal government that limit the value of any immunity. One should generally assume that the information developed as the result of an environmental audit is discoverable in civil litigation.

Failure to correct conditions found as the result of environmental audits carry risks of both civil and criminal liability. On the civil side, the existence of the environmental audit result can be used to demonstrate that there was knowledge of the condition likely to cause injury. On the criminal side the existence of the audit result can be used to demonstrate intent to violate environmental laws. These pitfalls indicate that environmental audits must be conducted with the involvement of counsel.

The final step in any environmental risk management program is monitoring. To successfully monitor a program there must be benchmarks established against which results can be assessed. If results fail to meet established benchmarks, the reason for such failures must be determined. Likewise, results that exceed benchmarks should also be assessed. Deficiencies should be corrected, and processes that work well should be applied more broadly as circumstances warrant.

No business ever completes the environmental risk management process; it is an iterative process. New risks arise, regulations and laws change, and existing risks that were previously unidentified come to light. Environmental risk management is a process of continuous improvement.

Adoption of a recognized environmental management system can be a very important part of an environmental risk management system. The EPA, as well as some state agencies, has been promoting adoption of environmental management systems (EPA 2004, http://www.epa.gov/ems/index.htm; North Carolina Department of Environment and Natural Resources 2004, http://www.p2pays.org/iso/). The ISO 14000 series of standards form the most widely recognized environmental management system. This system was developed by the International Organization for Standardization (International Organization for Standardization 2004, http://www.iso.ch/iso/en/ISOOnline.frontpage).

CHAPTER 7

# Occupational Safety and Health Administration Regulations and the Food Industry

Patricia Curtis, Auburn University

## Introduction

The Occupational Safety and Health Administration (OSHA) was created by Congress to help protect workers by setting and enforcing workplace safety and health standards and by providing safety and health information, training, and assistance to workers and employers. The Occupational Safety and Health Act of 1970 can be found on the OSHA website.[1]

Since the agency was created in 1971, workplace fatalities have been cut in half and occupational injury and illness rates have declined 40 percent. At the same time, U.S. employment has doubled from 56 million workers at 3.5 million worksites to 111 million workers at 7 million worksites. OSHA began fiscal year 2003 with a staff of 2,303, including 1,123 inspectors. Under the Bush administration, OSHA is focusing on three strategies: (1) strong, fair, and effective enforcement; (2) outreach, education, and compliance assistance; and (3) partnerships and voluntary programs. Each presidential administration has a slightly different focus.[2] OSHA's 2003–2008 strategic management plan goals are to reduce workplace fatality rates 15 percent and workplace injury and illness rates 20 percent by 2008.[3]

## Mission

The mission of OSHA is to save lives, prevent injuries, and protect the health of America's workers. To accomplish this, federal and state governments work in partnership with the more than 115 million working men and women and their 6.5 million employers who are covered by the Occupational Safety and Health Act of 1970.

The Occupational Safety and Health Act authorizes states to establish their own safety and health programs with OSHA approval. Twenty-three states operate state OSHA programs covering private sector workers, state government workers, and local government employees. In addition, Connecticut, New York, and New Jersey cover state and local government employees only. State OSHA programs must be at least as effective as the federal program and provide similar protections for workers. Some states set their own standards and others adopt federal rules. All state programs conduct inspections and respond to workers complaints and also provide other safety and health services, such as on-site consultation for small business.

## OSHA Statistics

Studies from occupational safety and health professionals show that women incur far fewer work injuries than men. In fact, Bureau of Labor Statistics data for the period 1992–1996 reveal that although they make up slightly under half of the total workforce, women incurred less than one-tenth of job-related fatal injuries and one-third of the nonfatal injuries and illnesses that required time off to recuperate.[4]

Immigrants come to the United States for a variety of reasons. Unfortunately, some immigrants find these dreams shattered as a result of violence, highway traffic accidents, or other fatal incidents in the workplace. Almost one-third of the foreign-born workers killed on the job in 1994 worked in retail trade, such as grocery stores and eating establishments, compared with one-eighth of the nation's native-born workers who died on the job that year. Most of these deaths were the result of being shot during an armed robbery attempt. By contrast, foreign-born workers who were fatally injured were less likely than their native cohorts to have worked in agricultural production, construction, mining, manufacturing, and public administration.[5]

Over the period from 1992 to 1997, 403 youths aged 17 years and under were killed on the job. These fatal incidents occurred primarily in agriculture, retail trade, construction, and services. One-third of the deaths occurred in family business and about one-half involved various types of vehicles and equipment.[6]

Results from a 1992 annual survey to profile the injury and illness experience of workers 55 years and older found older workers face a somewhat lower risk of sustaining a serious, nonfatal work disability than do younger workers. But older workers disabled on the job lose more work time per case than their younger counterparts. Older workers are also more likely to die of work-related injuries than their counterparts under age 55. In addition, the rates of fatal work injury rise with age, peaking at 15 deaths per 100,000 workers aged 65 years and older. In contrast, workers 25–34 years had a fatality rate of 5 per 100,000 in 1993.[7]

Fatalities resulting from workers being caught in machinery reached a six-year high in 1997. Half of the workers were performing service-related tasks at the time. Manufacturing is considered relatively safe in terms of fatal work injuries. In 1997, it accounted for 12 percent of fatal work injuries, compared with its 16 percent share of total employment. Agriculture, on the other hand, is considered relatively dangerous. In 1997, it accounted for 13 percent of the fatal work injuries, but comprised only 3 percent of total employment.[8]

In 2001, occupational injury and illness rates dropped to the lowest level—5.7 cases per 100 workers—since the United States began collecting this information. There were 5.2 million injuries/illnesses among private sector firms. There were 5,900 worker deaths in 2001. Fatalities related to highway incidents, electrocutions, fires and explosions, and contact with objects or equipment all declined. Deaths from job-related falls increased 10 percent, while homicides decreased to their lowest level since the census was first conducted in 1992. These figures do not include fatalities related to the events of September 11, 2001.

Workers under the influence of drugs or alcohol are a serious concern in the workplace today. Although the magnitude of the substance abuse problem at work is difficult to measure, the goal of maintaining alcohol- and drug-free workplaces has been addressed repeatedly by industry and government alike. In research conducted in 1993–1994 based on postmortem specimens, about one-fifth of the toxicology reports submitted showed positive readings for alcohol or one or more drugs. This is about 5 percent of total fatal work injuries.[9]

Violence has emerged as an important safety issue in today's workplace. Its most extreme form, homicide, is the second leading cause of death resulting from job-related injuries, accounting for 1,063 of 6,271 fatal injuries at work in 1993. Taxicab drivers and chauffeurs face unusually high risks of becoming homicide victims. Law enforcement and retail sales are other activities where the risks of homicide are especially high. Workers in retail establishments, such as convenience stores, retail groceries, and restaurants, face an above average risk. They account for about half of all homicides, but make up only one-sixth of the national work force.[10]

## Selected OSHA Standards and Guidelines

### Hazard Communications

More than 30 million workers in this country are exposed to hazardous chemicals in their work environment.[11] To protect these workers, OSHA adopted the Hazard Communication Standard (HCS) in November 1983. The standard requires chemical manufacturers and importers to evaluate the hazards of chemicals that they produce and distribute. The HCS requires information about hazards and protective measures to be disseminated on container labels and material safety data sheets (MSDSs). All employers with employees exposed to regulated chemicals must provide access to the labels and the MSDSs. Employers using the manufactured chemicals must also train their employees to understand the information provided by the MSDS and the labels and how to use the information to protect themselves.

The HCS covers all chemicals used in American workplaces. It is criteria based, so the standard is not limited to a list of chemicals at any given point in time. The standard addresses trade secrets to ensure protection of legitimate claims of confidentiality at the same time that it requires disclosure of safety and health information.

The HCS covers about 650,000 hazardous-chemical products in over three million work establishments. It has made the dissemination of hazard information about chemical products a standard business practice in the United States. There is now a generation of employers and employees who have continuously worked in an environment in which information about chemicals in their workplaces has been freely available.

MSDSs are the primary means of transmitting detailed chemical-hazard information to employers that use them and to their employees.

The MSDS is a technical bulletin, which contains information such as chemical composition, health hazards, and precautions for safe handling and use. Most safety and health professionals consider MSDSs to be a primary component of their company's hazard communication programs. Even prior to promulgation of the HCS, many chemical manufacturers and importers included MSDSs with hazardous chemicals as a good business practice.

The HCS places primary responsibility for preparing and disseminating the MSDSs with the chemical manufacturer. The HCS states clearly that manufacturers, importers, and employers preparing MSDSs shall ensure that the recorded information accurately reflects the scientific evidence used in making the hazard determination. However, MSDSs alone cannot protect workers from chemical hazards. The HCS also requires manufacturers to place labels on containers of hazardous chemicals and for employers using the manufactured chemicals to train their workforce.

## Ergonomics

A major component of OSHA's four-pronged approach to ergonomics is the development of industry-specific and task-specific guidelines to reduce and prevent workplace ergonomic injuries, often called musculoskeletal disorders (MSDs).[12] These voluntary guidelines are tools to assist employers in recognizing and controlling hazards. OSHA has voluntary guidelines published for meatpacking plants, the poultry industry, and retail grocery industry. OSHA plans to develop additional voluntary guidelines.

Ailments from performing repetitive tasks at work have been increasingly recognized and reported by physicians and employers. Federal government statistics on repeated-trauma disorders span a variety of ailments resulting from repeated motion, pressure, or vibration, such as carpal tunnel syndrome, tendonitis, and noise-induced hearing loss. Leading the list of industries with the largest number of repeated-trauma cases were motor vehicle and meat products manufacturing.

OSHA has developed recommendations for poultry-processing facilities to reduce the number and severity of work-related MSDs. The recommendations are based on a review of existing scientific literature and current programs and practices. The guidelines provided are for the poultry industry, but may benefit other industries. The focus of the recommendations is a description of various solutions that have been implemented in the poultry industry. OSHA states that the recommendations provided are advisory in nature and informational in content and they are not intended as a new standard or regulation.[13]

Employers should consider an MSD to be work-related if an event or exposure in the work environment either caused or contributed to the MSD, or significantly aggravated a preexisting MSD. For example, when an employee develops carpal tunnel syndrome and his or her job requires frequent hand activity or forceful or sustained awkward hand motions, then the problem may be work-related. If the job requires very little hand activity, then the disorder may not be work-related.

OSHA has a poultry processing industry eTool available online (http://www.osha.gov/SLTC/etools/poultry/evisceration.html). The eTool is an example of online training materials available from the OSHA website. The tool addresses specific tasks that may expose the worker to health and/or safety hazards. The eTool provides potential solutions and case studies of worker incidents.[14]

OSHA's injury and illness recording and reporting regulation (29 *CFR* 1904) requires employers to keep records of work-related injuries and illnesses. Employees may not be discriminated against for reporting a work-related injury or illness (29 *U.S.C.* 660(c)). OSHA recommends that employers implement a process that addresses the following areas:

1. Injury and illness recordkeeping
2. Early recognition and reporting
3. Systematic evaluation and referral
4. Conservative treatment
5. Conservative return to work (restricted duty)
6. Systematic monitoring
7. Adequate staffing and facilities.

Ergonomic solutions for the poultry industry have included engineering changes to work stations and equipment, work practices, personal protective equipment, and administrative actions. The poultry-processing industry has reduced occupational injuries and illnesses by almost half between 1992 and 2001, 23.2 per 100 full-time workers in 1992 to 12.7 in 2001.[21]

## Foodborne Disease

In situations where the employer is providing food to employees, certain standards must be met.[15] OSHA requires that potable water be provided in all places of employment, for drinking, washing of the employee, cooking, washing of foods, washing of cooking or eating utensils, washing of food preparation or processing premises, and personal service rooms (OSHA Standard 1910.141 (b), 1926.41 (a)). In addition, OSHA requires in all places of employment where all or part of the food service is provided, the food dispensed shall be wholesome, free from spoilage, and shall be processed, prepared, handled, and stored in such a manner as to be protected against contamination (OSHA Standard 1910.141(h), 1910.142 (i), 1926.51 (d)).

## Enforcement

OSHA's efforts to protect workers' safety and health are built on the foundation of a strong, fair, and effective enforcement program. OSHA will assist employers who want to do the right thing while focusing its enforcement resources on sites in high-hazard industries—especially those with high injury and illness rates.

In fiscal year 2002, OSHA conducted 37,493 inspections. Fifty-five percent of these inspections were high-hazard targets, 24 percent were complaint- or accident-related, and 21 percent were referrals or follow-ups. During this same time period, 58,402 state inspections were conducted. Sixty-one percent were high-hazard targets, 25 percent were complaint- or accident-related, and 14 percent were referrals or follow-ups.[2]

The ultimate measure of OSHA's effectiveness is the reduction of workplace injuries, illnesses, and fatalities. Records show OSHA is on the right track. The total recordable case rate has continued to decline. In addition to the decline in the rate of total recordable injuries and illnesses, the rate of cases that resulted in lost workdays also declined. The fatality rates have exhibited the same trend. Between 1998 and 2002, the number of fatalities decreased 8.3 percent.

Normally, OSHA conducts inspections without advance notice. Employers, however, have the right to require compliance officers to obtain an inspection warrant before entering the workplace. OSHA cannot inspect every workplace every year. They focus their inspection resources on the most hazardous workplaces. See Table 7.1 for priority categories.

For lower-priority hazards, OSHA may conduct a phone/fax investigation. OSHA may telephone the employer to describe the safety and/or health concern and follow up with a fax providing the details of the alleged hazards. The employer must respond in writing within five working

**Table 7.1.** OSHA priority categories

| *Priority* | *Category* | *Description* |
|---|---|---|
| 1 | Imminent danger situations | Hazards that could cause death or serious physical harm. Compliance officers will ask employers to correct these hazards immediately or will remove endangered employees |
| 2 | Fatalities and catastrophes | Incidents that involved a death or the hospitalization of three or more employees. Employers must report such catastrophes to OSHA within 8 hours. |
| 3 | Complaints | Allegations of hazards or violations. Employees may request anonymity when they file complaints. |
| 4 | Referrals | Referrals of hazard information from other federal, state, or local agencies, individuals, organizations, organizations or the media |
| 5 | Follow-ups | Checks of abatement of violations cited during previous inspections |
| 6 | Planned or programmed investigations | Inspections aimed at specific high-hazard industries or individual workplaces that have experienced high rates of injuries or illness |

days, identifying any problems found and describing corrective actions taken or planned. If the response is adequate and the complainant is satisfied with the response, OSHA will not conduct an on-site inspection.[16]

Before making an on-site inspection, the OSHA compliance officer researches the inspection history of a worksite using various data sources and then reviews the operations and processes in use and the standards most likely to apply. The inspector gathers appropriate personal protective equipment and testing instruments to measure potential hazards. Upon reaching the site, the inspector presents his or her credentials, explains why OSHA selected the workplace for inspection, and describes the scope of the inspection, walk-around procedures, employee representation, and employee interviews. The employer then selects a representative to accompany the compliance officer during the inspection. Following the opening conference, the compliance officer and the representative(s) will walk through portions of the workplace covered by the inspection, inspecting hazards that could lead to employee injury or illness. The officer will review worksite injury and illness records and posting of the official OSHA poster. During the walk-around, the compliance officer may point out some apparent violations that can be corrected immediately. After the walk-around, the compliance officer will hold a closing conference with the employer and the employee representative(s) to discuss the findings. The compliance officer will discuss possible courses of action an employer may take following an inspection, which could include an informal conference with OSHA or contesting citations and proposed penalties. OSHA must issue a citation and proposed penalty within six months of the violation's occurrence.

Citations describe OSHA requirements allegedly violated, list all proposed penalties, and give a deadline for correcting the alleged hazards. Violations are categorized as other than serious, serious, willful, repeated, and failure to abate. Penalties may range up to $700 for each serious violation and up to $70,000 for each willful or repeated violation. Penalties may be reduced based on employer's good faith, inspection history, and size of business.[17]

## Outreach, Education, and Compliance Assistance

OSHA plays a vital role in preventing on-the-job injuries and illnesses through outreach, education, and compliance assistance. OSHA offers an extensive website.[18] It includes a special section devoted to assisting small business as well as interactive eTools to help employers and employees. For example, the agency provides an interactive training program for lockout tagout (LOTO). Whether you are a recent hire or an experienced employee, this program will expand your knowledge of the (LOTO) standard. The program has three major components. You can go through these components at your own pace and in any sequence. The tutorial explains the standard in a question and answer format. Hot Topics contains five abstracts with a detailed discussion of major issues. Relevant highlighted sections of the all-inclusive documents are linked here. In interactive case studies, seven simulated LOTO inspections are presented. You will be making decisions on the application of the LOTO standard, based on information presented on the screen.[19]

OSHA provides a variety of publications in print and on CD-ROM, which are available from OSHA's regional or national offices or the Government Printing Office. OSHA tries to reach all employers and employees, including those who do not speak English as a first language. Many regional and area offices offer information in other languages such as Spanish, Japanese, Korean, and Polish.

Free workplace consultations are available in every state to small businesses that want on-site help in establishing safety and health programs and identifying and correcting workplace hazards. In addition, OSHA has a network of compliance assistance specialists in local offices available to provide tailored information and training to employers and employees.

OSHA's Strategic Partnership Program targets the strategic areas of construction, shipbuilding, food processing, logging, silica, and nursing homes, and includes partnerships that zero in on specific hazards or include partners in a specific geographic area. These partnerships focus on safety and health programs and include outreach and training components along with enforcement.

The Voluntary Protection Programs (VPPs) continues to pay big dividends. Today VPP worksites save millions each year because their injury and illness rates are more than 50 percent below the averages for their industries.

## Filing a Complaint with OSHA

The OSHA Act of 1970 gives employees the right to file complaints about workplace safety and health hazards.[20] Furthermore, the act gives complainants the right to request that their names not be revealed to their employers.[20] Complaints from employees and their representatives are taken seriously by OSHA. Complaints may be filed online. Most online complaints are addressed by OSHA's phone/fax system. That means complaints may be resolved informally over the phone with the employer. Written, signed complaints submitted to OSHA area or state plan offices are more likely to result in on-site OSHA inspections. Employees or their representatives must provide enough information for OSHA to determine that a hazard probably exists. Workers do not have to know whether a specific OSHA standard has been violated in order to file a complaint.

States with OSHA-approved state plans provide the same protections to workers as federal OSHA, although they may follow slightly different complaint-processing procedures. There are currently 23 states and jurisdictions operating OSHA-approved state occupational safety and health programs that cover both the private sector and state and local government authorities. Three other states operate approved state plans that cover state and local government employees only. Complaints to federal OSHA from workers in states with OSHA-approved state plans will be forwarded to the appropriate state plan for response.

## Summary

OSHA, as with all government organizations, has a website that offers the latest information available on a variety of topics. For the most current information on any OSHA-related topic check the *Federal Register* and the OSHA website.

## References

1. Occupational Safety and Health Act of 1970. Available at http://www.osha.gov/pls/oshaweb/owasrch.search_form?p_doc_type=OSHACT&p_toc_level=0&p_keyvalue= (Accessed 4/04).
2. Occupational Safety and Health Administration (OSHA). OSHA Facts. Available at http://www.osha.gov/as/opa/oshafacts.html (Accessed 5/04).
3. John L. Henshaw. 2003. OSHA's 2003–2008 Strategic Management Plan Goals: 15% Drop in Fatality Rates, 20% Drop in Injury and Illness Rates by 2008. American Industrial Hygiene Conference and Expo. May 12, 2003. Available at http://www.osha.gov/pls/oshaweb/owadisp.show_document?p_table=NEWS_RELEASES&p_id=10214 (Accessed 5/04).
4. Toscano, Guy A., Janice A. Windau, and Andrew Knestaut. 1998. Work Injuries Occurring to Women. *Compensation and Working Conditions*. Summer:16–22.
5. Windau, Janice. 1997. Occupational Fatalities Among the Immigrant Population. *Compensation and Working Conditions.* Spring:40–45.
6. Windau, Janice, Eric Sygnatur, and Guy Toscano. 1999. Profile of work injuries incurred by young workers. *Monthly Labor Review*. June:3–10.
7. Personick, Martin E. and Janice A. Windau. Characteristics of Older Workers' Injuries. Pp. 23–26.
8. Windau, Janice A. 1998. Worker Fatalities from Being Caught in Machinery. *Compensation and Working Conditions.* Winter:35–38.

9. Greenberg, Michael, Richard Hamilton, and Guy Toscano. 1999. Analysis of Toxicology Reports from the 1993–94 Census of Fatal Occupational Injuries. *Compensation and Working Conditions.* Fall: 26–28.
10. Tuscano, Guy and William Weber. Violence in the Workplace.
11. OSHA. 2004. OSHA's Hazard Communication Standard. Congressional Testimonies. March 25, 2004. Available at http://www.osha.gov/pls/oshaweb/owadisp.show_document?p_table=TESTIMONIES&p_id=349 (Accessed 4/04).
12. OSHA. Effective Ergonomics: Strategy for Success. A Four-Pronged, Comprehensive Approach. Available at http://www.osha.gov/SLTC/ergonomics/four-pronged_factsheet.html (Accessed 5/04).
13. OSHA. 2003. Ergonomics for Prevention of Musculoskeletal Disorders: Guidelines for Poultry Processing. Available at http://www.osha.gov/ergonomics/guidelines/poultryprocessing/index.html (Accessed 5/04).
14. OSHA. Poultry Processors eTool. Available at http://www.osha.gov/SLTC/etools/poultry/index.html (Accessed 5/04).
15. OSHA. 2003. Foodborne Disease: OSHA Standards. Available at http://www.osha.gov/SLTC/foodbornedisease/standards.html (Accessed 5/04).
16. OSHA Fact Sheet: OSHA Inspections. Available at http://www.osha.gov/OshDoc/data_General_Facts/factsheet-inspections.pdf (Accessed 4/04).
17. U.S. Department of Labor, OSHA. 2002. OSHA Inspections. OSHA 2098. Available at http://www.osha.gov/Publications/osha2098.pdf (Accessed 5/04).
18. OSHA website. Available at http://www.osha.gov (Accessed 5/04).
19. OSHA. Lockout/Tagout Tutorial: Control of Hazardous Energy. Available at http://www.osha.gov/dts/osta/lototraining/index.htm (Accessed 5/04).
20. OSHA. How to File a Complaint. Available at http://www.osha.gov/as/opa/worker/complain.html (Accessed 5/04).

## Additional Resources

Bureau of Labor Statistics. 1994. Shifting Work Force Spawns New Set of Hazardous Occupations. *Issues in Labor Statistics.* US Department of Labor, Summary 94–8, July.

Bureau of Labor Statistics. 1994. Repetitive Tasks Loosen Some Workers' Grip on Safety and Health. *Issues in Labor Statistics.* US Department of Labor, Summary 94–9, August.

Bureau of Labor Statistics. 1994. Violence in the Workplace Comes Under Closer Scrutiny. *Issues in Labor Statistics.* US Department of Labor, Summary 94–10, August.

Bureau of Labor Statistics. 2002. http://data.bls.gov/cgi-bin/dsrv?sh. Detailed occupational injury and illness industry data (1989–2001). Series ID SHU30201531 and SHU30201532, poultry processing.

Bureau of Labor Statistics. Census of Fatal Occupational Injuries: Definitions. http://www.bls.gov/iif/oshcfdef.htm.

Bureau of Labor Statistics. Frequently Asked Questions. http://www.bls.gov/iif/oshfaq1.htm.

Bureau of Labor Statistics. Occupational Safety and Health Definitions. http://www.bls.gov/iif/oshdef.htm.

Bureau of Labor Statistics. People Are Asking . . . . http://www.bls.gov/iif/peoplebox.htm.

Census of Fatal Occupational Injuries Staff. 1997. Perils in the Workplace. *Compensation and Working Conditions.* Fall:61–64.

Government Accountability Office. Workplace Safety and Health: OSHA's Voluntary Compliance Strategies Show Promising Results, but Should Be Fully Evaluated Before They Are Expanded. GAO-04-378 March 19, 2004. Available at http://www.gao.gov/docdblite/summary.php?recflag+&accno=A09507&rptno=GAO-04-378.

Older Workers' Injuries Entail Lengthy Absences From Work. 1996. *Issues in Labor Statistics.* U.S. Department of Labor, Bureau of Labor Statistics. Summary 96–6, April.

OSHA. Discrimination Against Employees Who Exercise Their Safety and Health Rights. http://www.osha.gov/as/opa/worker/whistle.html.

OSHA. Employee Responsibilities. http://www.osha.gov/as/opa/worker/responsible.html.

OSHA. Employer Responsibilities. http://www.osha.gov/as/opa/worker/employer-responsibility.html.

OSHA. Federal OSHA Complaint-Handling Process. http://www.osha.gov/as/opa/worker/handling.html.

OSHA. How to File a Complaint with OSHA. http://www.osha.gov/as/opa/worker/complain.html#what.

OSHA. OSHA Inspections. http://www.osha.gov/Publications/osha2098.pdf.

Toscano, Guy. 1997. Dangerous Jobs. *Compensation and Working Conditions*. Summer:57–60.

Toscano, Guy, Janice Windau, and Dino Drudi. 1996. Using the BLS Occupational Injury and Illness Classification System as a Safety and Health Management Tool. *Compensation and Working Conditions*, June:19–23.

CHAPTER 8

# Federal Trade Commission Regulations and the Food Industry

Patricia Curtis, Auburn University

The Federal Trade Commission (FTC) is an independent agency that reports to Congress on its actions. The FTC is headed by five commissioners, nominated by the president and confirmed by the Senate, each serving a seven-year term. The president chooses one commissioner to act as chair. No more than three commissioners may be from the same political party.[1]

## Mission

The FTC works to ensure that the nation's markets are vigorous, efficient, and free of restrictions that harm consumers. Experience demonstrates that competition among firms yield products at the lowest prices, spurs innovation, and strengthens the economy. Markets also work best when consumers can make informed choices based on accurate information.

To ensure the smooth operation of our free market system, the FTC enforces federal consumer protection laws that prevent fraud, deception, and unfair business practices. The FTC also enforces federal antitrust laws that prohibit anticompetitive mergers and other business practices that restrict competition and harm consumers. Whether combating telemarketing fraud, Internet scams, or price-fixing schemes, the FTC's primary mission is to protect consumers.

In addition, the commission conducts economic research and analysis to support its law enforcement efforts and to contribute to the policy deliberations of the Congress, the executive branch, other independent agencies, and state and local governments.[1]

## Authorizing Acts

The FTC deals with issues that impact the economic lives of most Americans. In fact, the agency has a long tradition of maintaining a competitive marketplace for both consumers and businesses. When the FTC was created in 1914, its purpose was to prevent unfair methods of competition in commerce. Over the years, Congress passed additional laws giving the agency greater authority to police anticompetitive practices.

In 1938, Congress passed the Wheeler-Lea Amendment, which included a broad prohibition against "unfair and deceptive acts or practices." Since then, the commission also has been directed to administer a wide variety of other consumer protection laws, including the Telemarketing Sales Rule, the Pay-Per-Call Rule, and the Equal Credit Opportunity Act.

In 1975, Congress passed the Magnuson-Moss Act, which gave the FTC the authority to adopt trade regulation rules that define unfair or deceptive acts in particular industries. Trade regulation rules have the force of law.

## Bureau of Consumer Protection

The Bureau of Consumer Protection's mandate is to protect consumers against unfair, deceptive, or fraudulent practices. The bureau enforces a variety of consumer protection laws enacted by Congress, as well as trade regulation rules issued by the commission. Its actions include individual company and industry-wide investigations, administrative and federal court litigation, rule-making proceedings, and consumer and business education. In addition, the bureau contributes to the commission's ongoing efforts to inform Congress and other government entities of the impact that proposed actions could have on consumers.

**Table 8.1.** Focus of enforcement activities for the Bureau of Consumer Protection's Division of Advertising Practices

- Claims for foods, drugs, dietary supplements, and other products promising health benefits
- Health fraud on the Internet
- Weight-loss advertising
- Advertising and marketing directed to children
- Claims about product performance made in national or regional newspapers and magazines; in radio and TV commercials, including infomercials; through direct mail to consumers; or on the Internet

The Bureau of Consumer Protection is divided into six divisions and programs, each with its own areas of expertise. Those divisions and programs related to food are discussed below.

## The Division of Advertising Practices

The Division of Advertising Practices is the nation's enforcer of federal truth-in-advertising laws. Its law enforcement activities related to food can be found in Table 8.1.

## The Division of Enforcement

The Division of Enforcement conducts a wide variety of law enforcement activities to protect consumers, including

1. ensuring compliance with administrative and federal court orders entered in consumer protection cases;
2. conducting investigations and prosecuting civil actions to stop fraudulent, unfair, or deceptive marketing and advertising practices; and
3. enforcing consumer protection laws, rules, and guidelines.

Enforcement of trade laws, rules, and guides through administrative or federal court proceedings include:

1. the Mail or Telephone Order Merchandise Rule, which requires "brick and mortar" and online companies to ship purchases when promised (or

within 30 days if no time is specified) or to give consumers the option to cancel their order for a refund; and

2. Green Guides, which govern claims that consumer products are environmentally safe, recycled, recyclable, ozone friendly, or biodegradable.

## Bureau of Competition

The FTC's antitrust arm, the Bureau of Competition, seeks to prevent anticompetitive mergers and other anticompetitive business practices in the marketplace. By protecting competition, the bureau promotes consumers' freedom to choose goods and services in an open marketplace at a price and quality that fit their needs. It also fosters opportunities for businesses by ensuring a level playing field among competitors.

The bureau fulfills this role by reviewing proposed mergers and other business practices for possible anticompetitive effects, and, when appropriate, recommending that the commission take formal law enforcement action to protect consumers. The bureau also serves as a research and policy resource on competition topics and provides guidance to business on complying with the antitrust laws.

### Research and Policy Studies

Congress created the FTC as a source of expertise and information on the economy. Consistent with this role, the Bureau regularly analyzes important competition-related topics.

## Bureau of Economics

The Bureau of Economics helps the FTC evaluate the economic impact of its actions. To do so, the bureau provides economic analysis and support to antitrust and consumer protection investigations and rulemakings. It also analyzes the impact of government regulation on competition and consumers and provides Congress, the executive branch, and the public with economic analysis of market processes as they relate to antitrust, consumer protection, and regulation.

## Sample Enforcement Actions

The following are a few examples of enforcement activities. The details for each activity are excerpts from FTC news releases.

### FTC Testifies on the Marketing of Dietary Supplements and Weight-Loss Products Containing Ephedra

"As the market for dietary supplements has expanded, so too have unfounded or exaggerated claims," said Howard Beales, Director of the Federal Trade Commission's Bureau of Consumer Protection.[2] In presenting testimony on July 24, 2003, before the House Committee on Energy and Commerce, Subcommittee on Oversight and Investigations, and the Subcommittee on Commerce, Trade, and Consumer Protection, Beales also noted that the commission's enforcement actions have expanded as well. "since December 2002, the Commission has targeted deceptive claims for more than $1 billion in health care products, a majority of which were dietary supplements," Beales stated.

Beales noted that the mission of the FTC is to prevent unfair competition and to protect consumers from unfair or deceptive practices in the marketplace. As part of this mission, the FTC has a long-standing and active program to combat fraudulent and deceptive advertising claims about the health benefits and safety of dietary supplements.

According to the testimony, ensuring the truthfulness of the advertising of health care products, and particularly dietary supplements, is a priority of the FTC's consumer protection agenda. During the past decade, the FTC had filed more than 90 law enforcement actions challenging false or unsubstantiated claims about the efficacy or safety of a wide variety of supplements. These enforcement actions include seven cases challenging claims for ephedra products marketed for weight loss, body building, and energy supplements, and as alternatives to street drugs such as Ecstasy. In each of these cases, the FTC imposed orders that prohibit unsubstantiated safety claims and require a strong disclosure warning about safety risks in all future advertising and labeling.

In September 2002, the FTC staff released a "Report on Weight-Loss Advertising: An Analysis of Current Trends," which recognized the detrimental effects of obesity and addressed the serious challenges facing law enforcement agencies in their efforts to stop deceptive advertising of weight-loss products and services. The report analyzed claims from 300 advertisements disseminated during 2001 and concluded that use of false or misleading weight-loss claims in advertising is widespread. The analysis showed that nearly 40 percent of the ads collected contained at least one representation that was almost certainly false, and 15 percent of the ads made at least one representation that was very likely to be false or to lack substantiation.

The testimony noted that ephedra is often marketed for use in weight loss. Beales testified that of the 300 advertisements sampled for the Weight Loss Advertising Report, 23, or about 8 percent, identified ephedra, ephedrine, or Ma Huang as an ingredient. Of these, 11 made safety claims, and 7 included a specific health warning about ephedra's potential adverse effects. Given that 60 percent of the sampled ads that made safety claims did not identify ingredients at all, these numbers almost certainly understate the prevalence of ephedra product advertising.

The testimony also noted that the FTC has met with members of the media and other interested parties to encourage them to screen out false weight-loss advertising claims.

In addition, Beales highlighted the FTC's longstanding liaison agreement with the Food and Drug Administration (FDA), and noted that since December 2002, the FTC and FDA have intensified their level of cooperation. "Our enforcement actions targeting false or unsubstantiated supplement safety claims play an important supporting role to the FDA's more comprehensive efforts to ensure the safety of supplement products," Beales said.

## Wonder Bread Marketers Settle FTC Charges

The marketers of Wonder Bread agreed to settle FTC charges that ads claiming that Wonder Bread containing added calcium could improve children's brain function and memory were unsubstantiated and violated federal law.[3] According to the FTC, the maker of Wonder Bread, Interstate Bakeries Corp. (IBC), aired an ad featuring a fictional spokesperson, Professor Wonder, who made claims that as a good source of calcium, Wonder Bread helped children's minds work better and helped their memory. The commission alleged that IBC and its ad agency did not have adequate substantiation to make such health benefit claims for Wonder Bread. The settlements announced March 6, 2002, barred the companies from making certain types of health benefit claims in the future, unless they have adequate substantiation.

According to the product's packaging, Wonder Bread is "fortified" and "enriched," containing several vitamins and minerals such as calcium, folic acid, iron, thiamine (vitamin B), riboflavin (vitamin B2), and niacin (vitamin B3). In the ad challenged by the FTC, Professor Wonder described certain purported health benefits of calcium—that it helped children's minds work better and helped memory—and emphasized that Wonder Bread has been fortified to be a "good source" of calcium. The complaints alleged that IBC did not have substantiation for the claims that, as a good source of calcium, Wonder Bread helps children's minds work better and helps them remember things. The Campbell complaint alleges that the ad agency knew or should have known these claims were unsubstantiated.

The settlements prohibited IBC from claiming that any bread product, or any of its ingredients, helped brain function or memory, or could treat, cure or prevent any disease or related health condition, unless they have reliable scientific substantiation for the claims. The orders allowed the respondents to make representations specifically permitted by the FDA.

Note that a consent agreement is for settlement purposes only and does not constitute an admission of a law violation. When the commission issues a consent order on a final basis, it carries the force of law with respect to future actions. Each violation of such an order may result in a civil penalty of $11,000.

## Bumble Bee Seafoods, Inc.

Nationally known as one of the major producers and sellers of canned tuna in the United States, Bumble Bee Seafoods, Inc. was based in San Diego, California. While the company markets many different types of tuna products that vary by size and content, the commission's complaint focused specifically on one coupon promotion that ran nationwide in 1998.[4]

In this promotion, the face of the can label indicated that consumers would find a coupon on the inside of the label enabling them to save seventy-five cents on their next Bumble Bee purchase. Because the coupon was printed on the inside of the label, consumers were only able to read the details of the offer once they had bought the product and removed the label at home. Once they did this, they found that they were required to purchase *five* cans of tuna in order to redeem the coupon. According to the FTC's complaint, the coupon campaign misled consumers into believing that they only had to purchase a *single* can of tuna to redeem the coupon.

Under the terms of the proposed agreement, Bumble Bee was prohibited from misrepresenting the terms or conditions of any rebate offer in the future and also would be required to disclose "clearly and prominently and in close proximity to the offer" the number of products that must be purchased in order for consumers to qualify for any rebate.

In addition, within 90 days after service of the order, Bumble Bee was required to begin a new consumer tear-pad coupon program, available in stores, that would include the national distribution of at least 7.6 million coupons. The coupons would prominently offer seventy-five cents off the purchase of "any two cans or multi-packs" of the company's solid white albacore tuna. The coupons would be redeemable at the place of purchase and would not expire until at least six months after being distributed.

The order further provided that if Bumble Bee's payout to consumers and incurred costs of the promotion do not exceed $200,000 within 90 days after the program ends, the company would be required to pay the difference between the actual cost and $200,000 to the U.S. Treasury. Finally, the order contained standard monitoring provisions to ensure the coupon program was put into place as specified and that the company kept all relevant records and made them available to the commission.

## FTC To Seek Injunction to Block Kroger Co. Purchase of Winn-Dixie Supermarkets in Texas and Oklahoma

Citing competitive concerns in Fort Worth and several smaller cities in Texas, the Federal Trade Commission voted to seek a preliminary injunction to block the Kroger Company's proposed acquisition of 74 Winn-Dixie supermarkets in Texas and Oklahoma. About half of the stores were in metropolitan Fort Worth, where Winn-Dixie and Kroger were the second- and third-largest supermarket chains, respectively.[5] According to the Commission complaint, the combined Kroger/Winn-Dixie presence in

Fort Worth accounted for 33 percent of all supermarket sales within the market, leading to the likelihood of competitive harm to consumers.

"The transaction as proposed would combine Fort Worth's second- and third-largest supermarket chains to create a new dominant firm in Fort Worth," said FTC's Bureau of Competition Director Richard G. Parker. "As a result of this merger, consumers would lose the benefit of more than 20 years of head-to-head competition between Kroger and Winn-Dixie."

Kroger, headquartered in Cincinnati, Ohio, was the largest supermarket chain in the United States, operating more than 2,200 stores under the "Kroger" banner and several others in 31 states. Its U.S. sales in 1999 were more than $45 billion. Kroger operated 54 supermarkets in Texas, with average weekly sales in metropolitan Fort Worth totaling $6.2 million.

Winn-Dixie Texas, Inc., a Texas corporation and wholly owned subsidiary of Winn-Dixie, was headquartered in Fort Worth. It operated 74 supermarkets in Texas and Oklahoma. In these states, Winn-Dixie's total sales revenue for fiscal year 1999 was $745 million, with average sales of $6.8 million per week in metropolitan Forth Worth.

According to the commission's complaint, Kroger's acquisition of Winn-Dixie's stores in Texas and Oklahoma would violate Section 7 of the Clayton Act, 15 *U.S.C.* § 18, and Section 5 of the Federal Trade Commission Act, 15 *U.S.C.* § 45, by substantially reducing competition in several markets in Texas including metropolitan Fort Worth, Granbury, Weatherford, Brownwood, Henderson, Denton, and Marshall.

While Dallas and Fort Worth are within an area known as the Metroplex, in this case the commission defined metropolitan Fort Worth as a market economically and geographically distinct from Dallas for supermarket consumers. Within the Fort Worth market, if the acquisition were to proceed as proposed, Kroger would control one-third of all supermarket sales within the city. According to the complaint, the smaller markets outside Fort Worth were even more highly concentrated, with the merged firm poised to become the largest chain in Granbury, Weatherford, Brownwood and Denton, and the second-largest in Marshall and Henderson.

Additional evidence suggested that the acquisition would harm competition by allowing Kroger to exercise unilateral market power, according to the commission. First, the acquisition would end 22 years of direct competition between the two supermarket chains, enabling Kroger to surpass Albertsons and become the "market leader" in Fort Worth. While Winn-Dixie was Krogers' main competitor in this market, research indicates that most Winn-Dixie customers view Kroger as a close substitute when doing their food shopping. It was expected, therefore, that they would simply switch to Kroger if the Winn-Dixie where they currently shop were acquired and renamed. Maintaining the Winn-Dixie customer base would help Kroger increase its already strong position in the marketplace. Also, some Kroger supermarkets were located so close to Winn-Dixie's that Kroger could close specific Winn-Dixie's that "overlap" because of their proximity to the Kroger-branded outlets. Kroger could even close both stores, forcing its customers to shop at a nearby supermarket. In either case, due to the close customer association between the two chains, most customers would most likely become or remain Kroger shoppers.

Lastly, the acquisition would increase the chance that Kroger could engage in anticompetitive interactions with the remaining supermarket chains, such as Albertson's, Tom Thumb, and Minyards. By eliminating the direct competition between Kroger and Winn-Dixie, it would also eliminate the need for future competition between the two chains, both of which appear to have aggressive growth strategies in the Fort Worth area. The complaint also alleged that new stores operated by competing firms were not expected to be opened in sufficient numbers in these markets to defeat Kroger's ability to exercise market power after the transaction is completed.

The commission voted to seek a injunction to block the transaction.

## World's Largest Manufacturer of Spice and Seasoning Products Agrees to Settle Price Discrimination Charges

McCormick & Company, the world's largest spice company, has agreed to settle Federal Trade Commission charges that it violated federal antitrust laws by engaging in unlawful price discrimination in the sale of its spice and seasoning products.[6] According to the FTC, for a substantial period of time, McCormick charged some retailers a substantially higher net price for its spice and seasoning products than it charged other competing retailers. The FTC alleged that McCormick sold its products at different prices by providing competing retailers discriminatory aggregate discounts off the list prices of its products. These aggregate discounts, known in the industry as "deal rates," took a variety of forms, including up-front cash payments similar to slotting allowances, free goods, off-invoice discounts, cash rebates, performance funds, and other financial benefits. According to the complaint, the victims of McCormick's price discrimination, referred to in the complaint as "disfavored purchasers," had few, if any, alternative sources from which to purchase comparable goods at prices and terms equivalent to those which McCormick provided to the favored purchasers. The settlement resolved claims arising under the Robinson-Patman Act, which prohibited sellers from charging competing buyers different prices for goods of "like grade and quality," where "the effect of such discrimination may be substantially to lessen competition or tend to create a monopoly in any line of commerce . . . ." The order prohibited McCormick from engaging in price discrimination that violates the Act, unless the price differences are permitted by defenses recognized by the Act.

According to Richard Parker, Director of the FTC's Bureau of Competition, "The Order sends a strong message to dominant sellers that the Commission will not tolerate price discrimination where buyers have little choice but to do business with the discriminating company."

Maryland-based McCormick & Company was the largest manufacturer of spice and seasoning products in the world. It had 1998 retail sales of $623.7 million in the Americas. McCormick sells products for retail sale under a number of brand names, including McCormick, Schilling, Fifth Seasons, Spice Classics, Select Seasons, Mojave, Spice Trend, Royal Trading, Crescent, La Cochina De McCormick, and Old Bay. The FTC alleged that McCormick typically required that its customers allocate the large majority of the spice shelf to McCormick. In some instances, this requirement covered 90 percent of the supermarket's spice shelf space. According to the FTC, among firms supplying core or gourmet spice lines for sale in supermarkets in the United States, McCormick was by far the leading firm, accounting for the majority of such sales nationally. During the period pertinent to this complaint, McCormick faced competition in such sales from only one other national firm, Burns Philp Food Incorporated, and several, much smaller independent regional or local firms.

The consent agreement prohibited McCormick from selling its products to any purchaser at a net price higher than McCormick charged the purchaser's competitors, except when permitted by the Robinson-Patman Act. "Net price" was defined in the order as the list price of McCormick products less advances, allowances, discounts, rebates, deductions, free goods and other financial benefits provided by McCormick and related to such products. The order applied to McCormick's sale of spices, seasonings, and other products which were used to season or flavor foods, packaged by McCormick for resale to consumers.

The order permitted McCormick to sell its products to competing purchasers at different prices when such price differentials were permitted by the statutory "meeting competition" defense. The "meeting competition" defense under the Robinson-Patman Act allows a seller to provide a lower price to a purchaser when the seller believes that it must provide the lower price to meet the equally low price offered by its competitor. If the lower price is offered to meet the seller's competition, the seller is not required to provide the same low price to the buyer's competitors. The proposed consent order goes beyond the statutory requirements of the Act by requiring McCormick for five years, for each instance in which it wishes to avail itself of the "meeting competition" defense, to contemporaneously document and maintain in its records all information on which it bases its entitlement to the defense. This information would assist the FTC in assuring that McCormick complied with the order and with the Act.

In addition, the order, which was effective for 20 years, required McCormick to distribute a copy of the order to all of its current and new employees who were involved in the sale of products covered by the order and also required McCormick to inform the FTC of corporate changes that may affect its compliance obligations. McCormick also would be required to file reports demonstrating its compliance with the order.

## Seven Internet Retailers Settle FTC Charges Over Shipping Delays During 1999 Holiday Season

Seven large Internet e-tailers agreed to settle Federal Trade Commission charges they violated the Mail and Telephone Order Rule during the 1999 holiday shopping season by providing buyers inadequate notice of shipping delays or continuing to promise specific delivery dates when timely fulfillment was impossible. In settling the allegations, CDnow, Inc., KBkids.com LLC, Macys.com, Inc., Franklin W. Bishop d/b/a Minidiscnow.com, The Original Honey Baked Ham Company of Georgia, Inc., Patriot Computer Corp., and Toysrus.com, Inc. have agreed to change their procedures to ensure that such violations will not recur and to pay civil penalties totaling $1.5 million violations.[7] The settlements were the culmination of "Project TooLate.com," an FTC investigation of whether major online retailers delivered goods when promised during the holiday season.

The requirements of the Rule apply to online and offline commerce equally; this settlement demonstrated that the FTC takes violations of the

Rule by e-tailers seriously and will expect e-tailers either to comply with the law or face stiff penalties.

The FTC received complaints from a variety of sources indicating that prominent e-tailers repeatedly failed to meet their express shipping and delivery representations.

Based on this information, FTC staff began an investigation of e-tailers, focusing on major online companies that were making express delivery claims, to examine whether these failures were violations of the Rule.

The Rule requires that retailers ship goods within the date promised, or if no date is promised, within 30 days of the order's receipt. If the company cannot ship as promised, it is required to provide Notice to the buyer with a revised shipping date, giving the opportunity for the buyer to agree to the delay or to cancel the order.

The staff investigation revealed that some e-tailers repeatedly violated the Rule by failing to send Rule-required Delay Option Notices, sending Notices that were deficient, and in some cases, making shipping representations without a reasonable basis.

The FTC obtained settlements from seven companies which it alleged violated the Rule. Common to all complaints was the failure to timely offer to the buyer an option either to consent to a delay in shipping or to cancel the order and receive a prompt refund, in violation of Section 435.1(b)(1) of the Rule.

Six companies (all but CDnow) that sent no Notices were charged with failure to deem certain orders canceled in the absence of a notice, in violation of Section 435.1(c) of the Rule. Four defendants (KBkids.com, Macys.com, Toysrus.com, and Minidiscnow.com) were charged with taking orders without a reasonable basis for their shipping representations, in violation of Section 435.1(a)(1) of the Rule.

The consent decrees with six of the companies called for civil penalties ranging from $45,000 to $350,000—Macys.com ($350,000), KBkids.com ($350,000), Toysrus.com ($350,000), CDnow ($300,000), Patriot Computers ($200,000), and The Original Honey Baked Ham Co. ($45,000). In addition, Macys.com was required to fund an Internet consumer education campaign about the Mail Order Rule. The campaign consisted of banner ads that alerted consumers that they have certain rights when shopping online. The penalty amounts sought against KBkids.com, Toysrus.com, and The Original Honey Baked Ham Co. took into account the amount of money the companies spent in mitigating consumer injury caused by their Rule violations. CDnow's penalty amount was waived except for

$100,000 due to its poor financial condition. The seventh company, Minidiscnow.com, was required to fully reimburse each consumer who ordered, but did not receive, any of the company's products.

## Two Vending Machine Franchisors Pay Civil Penalty To Settle FTC Charges

American Coin-Op Services, Inc., based in Buffalo, New York, and its principals, Daryl J. Chase and Craig J. Schieder, agreed to settle Federal Trade Commission charges that they failed to provide the pre-sale disclosures required by the FTC's Franchise Rule to prospective purchasers of their mint and candy vending machine opportunities.[8] In a separate settlement, World Wide Coffee, Inc., based in Jupiter, Florida, and its principals, Jeffrey Salley and Terri Salley, also agreed to settle FTC charges that they failed to provide the pre-sale disclosures to prospective purchasers of their espresso coffee machine vending opportunities.

The Department of Justice, at the request of FTC, filed suits against the two companies and their principals as part of "Project Biz-illion$," a nationwide crackdown on fraudulent business opportunities. Under the terms of both settlements, the defendants were prohibited from violating the Franchise Rule and making false and misleading representations in connection with the sale of business opportunities. In addition, the American Coin-Op Services defendants agreed to pay an $11,000 civil penalty, and the World Wide Coffee defendants agreed to pay a $20,000 civil penalty.

These settlements ended the litigation in these cases, which were among 22 cases the FTC referred to the DOJ (Department of Justice) for filing as part of "Project Biz-illion$," a multi-prong state/federal attack on business opportunity scams. These cases, like most of the "Project Biz-illion$" actions, were launched against defendants that advertised in the classified section of daily newspapers to peddle payphone, vending machine, display rack, and work-at-home scams.

According to the FTC, the defendants in the *American Coin-Op Services* case represented that they would set purchasers up in their own profitable vending business for a minimum investment of $5,985 for 15 vending machines. The defendants' promotional materials represented that prospective purchasers could expect to earn annual profits from $10,912 with 20 machines to $102,300 with 100 machines, and their classified ads claimed the business had a "$1000/wk potential."

Similarly, the defendants in the *World Wide Coffee* case represented that they would furnish the equipment and coffee, and secure restaurant and bar locations for their espresso coffee vending opportunity for a minimum investment of $9,300 for a package of five espresso machines. The defendants' promotional materials represented that prospective purchasers could expect to earn an income of $15,000 with 5 machines averaging 25 servings a day, and their classified ads claimed the business would earn "2K/wkly."

In both cases, the complaints alleged that the defendants failed to provide prospective purchasers with an earnings claim document providing written substantiation for the defendants' earnings claims, including the number and percentage of prior purchasers who had earned that much, as the Franchise Rule requires. The complaints also alleged that the defendants failed to provide prospective purchasers with a basic disclosure document that included the names, addresses and telephone numbers of prior purchasers, as required by the Rule, to help potential purchasers protect themselves from false profitability claims.

In addition to paying civil penalties, the settlements, which required the court's approval, prohibited the defendants from future violations of the Franchise Rule; making false and misleading representations in connection with the sale of business opportunities; and selling their customer lists. The settlements also contained various recordkeeping and reporting requirements designed to assist the FTC in monitoring the defendants' compliance.

## References

1. FTC. 2004. Guide to the Federal Trade Commission. Available at http://www.ftc.gov/bcp/conline/pubs/general/guidetoftc.pdf (Accessed 8/04).
2. FTC. 2003. FTC Testifies on the Marketing of Dietary Supplements and Weight-Loss Products Containing Ephedra. News Release (July 24, 2003). Available at http://www.ftc.gov/opa/2003/07/ephedratestimony.htm (Accessed August 23, 2004).
3. FTC. 2002. Wonder Bread Marketers Settle FTC Charges. News Release (March 6, 2002). Available at http://www.ftc.gov/opa/2002/03/wonderbread.htm (Accessed August 23, 2004).
4. FTC. 2000. Settlements Reached with Bumble Bee Seafoods, Memtek Products and UMAX Technologies. FTC News Release (January 10,

2000). Available at http://www.ftc.gov/opa/2000/01/bumblebee.htm (Accessed August 23, 2004).
5. FTC. 2000. FTC to Seek Injunction to Block Kroger Co. Purchase of Winn-Dixie Supermarkets in Texas and Oklahoma. FTC News Release (June 2, 2000). Available at http://www.ftc.gov/opa/2000/06/krogerwinndixie.htm (Accessed August 23, 2004).
6. FTC. 2000. World's Largest Manufacturer of Spice and Seasoning Products Agrees to Settle Price Discrimination Charges. FTC News Release (March 8, 2000). Available at http://www.ftc.gov/opa/2000/03/mccormick.htm (Accessed August 23, 2004).
7. FTC. 2000. Seven Internet Retailers Settle FTC Charges Over Shipping Delays During 1999 Holiday Season. FTC News Release (July 26, 2000). Available at http://www.ftc.gov/opa/2000/07/toolate.htm (Accessed August 23, 2004).
8. FTC. 2001. Two Vending Machine Franchisors Pay Civil Penalty To Settle FTC Charges. FTC News Release (April 17, 2001). Available at http://www.ftc.gov/opa/2001/04/amcoinop.htm (Accessed August 23, 2004).

CHAPTER 9

# An Introduction to Kosher and Halal Food Laws*,**

Joe M. Regenstein, Professor of Food Science, Cornell Kosher Food Initiative, Department of Food Science Cornell University
Muhammad M. Chaudry, Executive Director, Islamic Food and Nutrition Council
Carrie E. Regenstein, Associate Chief Information Officer Associate Director, Division of Information Technology

## Introduction

The objective of this chapter is to describe the kosher and halal laws as they apply in the food industry, particularly the United States, and to understand how secular regulatory authorities ensure the integrity of the process. To understand their impact in the marketplace and the courtroom one must have some understanding of how kosher and halal foods are produced, and how important kosher and halal compliance is to consumers.

### The Kosher and Halal Laws

We will start by focusing on the religious significance of the dietary laws for Jews and Muslims. The kosher (kashrus) dietary laws determine

*We thank Dr. Shawkat Toorawa of Cornell University's Near Eastern Studies Department for critical comments.

**The information in this chapter is as accurate as possible (as of November 1, 2003). However, the final decision on any product application rests with the religious authorities providing supervision and the interpretation of secular law is that of the authors. The rulings of the religious and secular authorities may differ from the information presented here.

which foods are "fit or proper" for consumption by Jewish consumers who observe these laws. The laws are biblical in origin, coming mainly from the original five books of the Holy Scriptures, the Torah, which has remained unchanged. At the same time that Moses received the Ten Commandments on Mount Sinai, Jewish tradition teaches that he also received the oral law, which was eventually written down many years later in the Talmud. This oral law is as much a part of biblical law as the written text. Over the subsequent years, the meaning of the biblical kosher laws have been interpreted and extended by the rabbis to protect the Jewish people from violating any of the fundamental laws and to address new issues and technologies. The system of Jewish law is referred to as *halacha*. It is a legal system with both similarities to and differences with modern law in Western society.

The halal dietary laws determine which foods are lawful or permitted for Muslims. These laws are found in the Quran and in the Sunna, the practice of the Prophet Muhammad, as recorded in the books of Hadith, the Traditions. Islamic law is referred to as *Shari'ah* and has been interpreted by Muslim scholars over the years. The basic principles of the Islamic laws remain definite and unaltered. However, their interpretation and application may change according to the time, place, and the circumstances. Besides the two basic sources of Islamic law, The Quran and the Sunna, two other systems of jurisprudence are used in determining the permissibility of food when a contemporary situation is not explicitly covered by the first two basic sources. The first is *Ijma*, meaning a consensus of legal opinion. The second is *Qiyas*, meaning reasoning by analogy. In the latter case the process of *Ijtihad*, or exerting oneself fully to derive an answer to the problem, is used. Again, a system of religious law exists, of which the food laws are just a small part, and may differ significantly from Western legal systems.

Why do Jews follow the kosher dietary laws? Many explanations have been given. The explanation by Rabbi I. Grunfeld summarizes the most widely held ideas about the subject (Grunfeld, 1972). Notice that it emphasizes the importance of the legal structure.

It is important to note that, unlike the kosher laws, the health aspects of eating are an important consideration with the halal laws. These laws are viewed by the Jewish community as given to the community without a need for explanation. Only in modern times have some people felt a need to try to justify them as health laws. For a discussion of why the kosher laws are not health laws, please see J.M. Regenstein, 1994.

> "And ye shall be men of a holy calling unto Me, and ye shall not eat any meat that is torn in the field" (Exodus XXII:30). Holiness or self-sanctification is a moral term; it is identical with . . . moral freedom or moral autonomy. Its aim is the complete self-mastery of man.
>
> To the superficial observer it seems that men who do not obey the law are freer than law-abiding men, because they can follow their own inclinations. In reality, however, such men are subject to the most cruel bondage; they are slaves of their own instincts, impulses and desires. The first step towards emancipation from the tyranny of animal inclinations in man is, therefore, a voluntary submission to the moral law. The constraint of law is the beginning of human freedom. . . . Thus the fundamental idea of Jewish ethics, holiness, is inseparably connected with the idea of Law; and the dietary laws occupy a central position in that system of moral discipline which is the basis of all Jewish laws.
>
> The three strongest natural instincts in man are the impulses of food, sex, and acquisition. Judaism does not aim at the destruction of these impulses, but at their control and indeed their sanctification. It is the law which spiritualizes these instincts and transfigures them into legitimate joys of life.

Why do Muslims follow the halal dietary laws? The main reason for the observance of the Islamic faith is to follow the Divine Orders.

> O ye who believe! Eat of the good things wherewith WE have provided you, and render thanks to ALLAH if it is He whom ye worship. (Quran II:172)

G-d reminds the believers time and again in the Holy Scripture to eat what is *Halalan Tayyiban*, meaning permitted and good or wholesome.

> O, Mankind! Eat of that which is Lawful and Wholesome in the earth . . . (Quran II:168)
>
> Eat of the good things. We have provided for your sustenance, but commit no excess therein. (Quran XX:81)

Again in Sura 6 of the Quran, "Cattle," Muslims are instructed to eat the meat of animals upon which Allah's name has been invoked. This is generally interpreted as meaning that an invocation has to be made at the time of slaughtering an animal.

> Eat of that over which the name of Allah hath been mentioned, if ye are believers in His revelations. (Quran VI:119)

While Muslims eat what is permitted specifically or by implication, albeit without comment, they avoid eating what is specifically disallowed, such as:

> And eat not of that whereupon Allah's name hath not been mentioned, for lo, it is abomination. Lo! The devils do inspire their minions to dispute with you. But if ye obey them, ye will be in truth idolaters. (Quran VI:121)

The majority of the Islamic scholars are of the opinion that this verse deals with proper slaughtering of the allowed animals.

Since Muslim dietary laws relate to divine permissions and prohibitions, if anyone observes these laws, he/she is rewarded in the hereafter, and if anyone violates these laws, he/she may receive punishment accordingly. The rules for those foods that are not specifically prohibited may be interpreted differently by various scholars. The things that are specifically prohibited are just a few in number, and are summarized in the following verses:

> Forbidden unto you are: carrion and blood and swine flesh, and that which hath been dedicated unto any other than Allah, and the strangled, and the dead through beating, and the dead through falling from a height, and that which hath been killed by the goring of horns, and the devoured of wild beasts save that which ye make lawful, and that which hath been immolated to idols. And that ye swear by the divining arrows. This is abomination. (Quran V:3)

Although these permissions and prohibitions as a divine injunction are enough for a Muslim to observe the laws, it is believed that the dietary laws are based on health reasons that suggest impurity or harmfulness of prohibited foods.

## The Kosher and Halal Market

Why are we concerned about kosher and halal in the secular world? Because both kosher and halal are important components of the food business. Most people, even in the food industry, are not aware of the breadth of foods that are under religious supervision. This section provides a brief background on the economic aspects that make it important for the food industry to have a better understanding of kosher and halal.

The kosher market, according to Integrated Marketing, an advertising agency specializing in the kosher food industry, comprises almost 75,000 products in the United States. In 2001, about $165 billion worth of products are estimated to have a kosher marking on them. The deliberate consumers of kosher food, that is, those who specifically look for the kosher mark, are estimated to be over 10 million Americans, and they are purchasing almost $7 billion worth of kosher products. Annually, almost 10,000 companies produce kosher products, and the average U.S. supermarket has 13,000 kosher products. Less than one-third (possibly as low as 20 percent) of the kosher consumers are Jewish (900,000 year-round consumers). Other consumers, who at times find kosher products helpful in meeting their dietary needs, include Muslims, Seventh Day Adventists, vegetarians, vegans, people with various types of allergies—particularly to dairy, grains, and legumes—and general consumers who value the quality of kosher products, even though there is rarely a one-to-one correlation between kosher and these consumers' needs. By undertaking kosher certification, companies can incrementally expand their market by opening up new markets. This chapter also includes information that might assist kosher supervision agencies in addressing the specific needs of these other consumer groups and regulatory agencies in understanding some of the issues that need to be addressed in assuring the integrity of the marketing process.

The Muslim population in the United States is developing a stronger marketplace presence each year. Over the past 30 years many halal markets and ethnic stores have sprung up, mainly in the major metropolitan areas. Most of the six to eight million Muslims in North America observe halal laws, particularly the avoidance of pork, but the food industry has, for the most part, ignored this consumer group. Although there are excellent opportunities to be realized in the North American halal market, even more compelling opportunities exist on a worldwide basis as the food industry moves to a more global business model. The number of Muslims in the world is over 1.3 billion people, and trade in halal products is about 150 billion dollars (Egan 2002). Many countries of South Asia, Southeast Asia, the Middle East, and northern Africa have predominantly Muslim populations. In many countries halal certification has become necessary for products to be imported.

Although many Muslims purchase kosher food in the United States, these foods, as we will see in the section on halal, do not always meet the needs of the Muslim consumer. The most common areas of concern for the Muslim consumer when considering purchasing kosher products are the use of various questionable gelatins in products produced by more lenient kosher supervisions and the use of alcohol as a carrier for flavors, as well as a food ingredient. The details of both ideas will be developed later in this chapter.

With the agreement of the client company, kosher supervisors can address the needs of the non-Jewish markets. A document establishing preliminary guidelines for making kosher appropriate for all of the groups mentioned above without violating Jewish law has been prepared (Regenstein, personal communication).

# Kosher

## The Kosher Dietary Laws

The kosher dietary laws predominantly deal with three issues, all focused on the animal kingdom:

1. Allowed animals
2. Prohibition of blood
3. Prohibition of mixing of milk and meat

Additionally, for the week of Passover (in late March or April) restrictions on *chometz*, the prohibited grains (wheat, rye, oats, barley, and spelt) in other than unleavened form, and the rabbinical extensions of this prohibition, lead to a whole new set of regulations, focused in this case on the plant kingdom.

Ninety-two percent of American Jews celebrate Passover in some way, making it the most observed holiday in the Jewish calendar. It also accounts for about 40 percent of the sales of kosher products to the Jewish community. Although only 20 to 33 percent of the kosher market in the United States is Jewish, these consumers account for over half of the total dollar volume of the kosher market, since they purchase kosher food more consistently.

We will also discuss additional laws dealing with special issues such as grape juice, wine, and alcohol derived from grape products; Jewish supervision of milk; Jewish cooking, cheese making, and baking; equipment kosherization; purchasing new equipment from non-Jews; and old and new flour.

The kosher laws are an internally consistent logic system and have an implied "science" behind them, which may or may not agree with modern science. This system is the basis upon which rabbis work through problems and come up with solutions. It is a system of case law reflected in written (and oral) response, essentially the equivalent of a judicial opinion.

### Allowed Animals

Ruminants with split hoofs that chew their cud, the traditional domestic birds, and fish with fins and removable scales are generally permitted. Pigs, wild birds, sharks, dogfish, catfish, monkfish, and similar species are prohibited, as are all crustacean and molluscan shellfish. Almost all insects are prohibited, such as carmine and cochineal, which are used as natural red pigments, are not permitted in kosher products by most rabbinical supervisors. However, honey and shellac (lac resin) are permitted, as will be discussed later in this section.

Four classes of prohibited animals are specifically described in the Torah. These are those animals that have one kosher characteristic but not both. The rockbadger, the hare, and the camel chew their cud but do not have a split hoof; the pig has a split hoof but does not chew its cud. Neither those with one characteristic nor those with none are kosher. In modern times, the prohibition of pork has often been the focus of both kosher and halal laws, since pork is such a major item of commerce. With respect to poultry, the traditional domestic birds, that is, chicken, turkey, squab, duck, and goose are kosher. Birds in the rattrie category (ostrich, emu, and rhea) are not kosher, as the ostrich is specifically mentioned in the Bible (Lev. XI:16). However, it is not clear whether the animal of the Bible is the same animal we know today as an ostrich.

The only animals from the sea that are permitted are those with fins and scales. All fish with scales have fins, so the focus is on the scales. These must be visible to the human eye and must be removable from the fish skin without tearing the skin. Cycloid and ctenoid scales found on traditional fish are generally considered acceptable, but the ganoid and placoid scales of sharks, gar, etc., are not. A few fish remain controversial, probably swordfish, whose scales do not seem to belong to any of the biologists standard scale types, being the most discussed fish. The Conservative movement permits it along with sturgeon, while most Orthodox authorities consider both non-kosher.

Most insects are not kosher. The exception includes a few types of grasshoppers, which are acceptable in the parts of the world where the tradition of eating them has not been lost. Only visible insects are of concern;

an insect that spends its entire life cycle inside a single food is not of concern. And, the prohibition of insects focuses on the whole animal. Modern IPM (integrated pest management) programs that increase the level of insect infestation in fruits and vegetables can cause problems for the kosher consumer.

Honey and other products from bees are covered by a unique set of laws that essentially permits honey and beeswax. Other bee derived materials, for example, royal jelly, are more controversial. An article by Rabbi Z. Blech (2004) discusses this unique set of materials and the special laws surrounding bees and honey. Most rabbis extend this permission to the use of lac resin or shellac, which is used in candy and fruit coatings to provide a shine.

### Prohibition of Blood

Ruminants and fowl must be slaughtered according to Jewish law by a specially trained religious slaughterman (*shochet*) using a special knife designed for the purpose (*chalef*). The knife must be extremely sharp and have a very straight blade that is at least twice the diameter of the neck of the animal to be slaughtered. It is the process itself, and the strict following of the law, that makes a product kosher, and not the presence or absence of a blessing over the food. However, prior to slaughter the shochet does make a blessing. The animal is not stunned prior to slaughter. If the slaughter is done in accordance with Jewish law and with the highest standards of modern humane animal welfare handling practices, the animal will die without showing any signs of stress. In 1958, the U.S. Congress declared kosher slaughter and similar systems (e.g., such as halal) to be humane, but included an exemption for preslaughter handling of the animal prior to kosher and halal slaughter. To deal with problems due to inappropriate preslaughter handling, the Food Marketing Institute (FMI), the trade association for many North American supermarkets, and the National Council of Chain Restaurants (NCCR) are developing a set of animal welfare–based kosher/halal standards for upright slaughter based on the American Meat Institutes' guidelines that have existed for a number of years as part of a program to develop a modern set of animal welfare guidelines without involving the federal government directly.

With respect to kosher, or kashrus, supervision, slaughtering is the only time a blessing is said—and it is said before commencing slaughter. The slaughterman asks forgiveness for taking a life. The blessing is not said over each animal, an issue we will return to when discussing the Muslim concept of the meat of the "People of the Book."

Slaughtered animals are subsequently inspected for visible internal organ defects by rabbinically trained inspectors. If an animal is found to have a defect, the animal is deemed unacceptable and becomes *treife*. There is no trimming of defective portions as generally permitted under secular law.

Consumer desire for more stringent kosher meat inspection requirements in the United States has led to the development of a standard for kosher meat that meets a stricter inspection requirement, mainly with respect to the condition of the animal's lungs. As the major site of halachic defects, the lungs must always be inspected. Other organs are spot-checked or examined when a potential problem is observed. Meat that meets this stricter standard is referred to as *glatt kosher*, referring to the fact that the animal's lungs do not have any adhesions (*sirkas*). The word *glatt* means smooth, referring to the absence of sirkas on the lungs. The *bodek*, or the inspector of the internal organs, is trained to look for lung adhesions in the animal both before and after its lungs are removed. To test a lung, the bodek first removes all sirkas and then blows up the lung using normal human air pressure or a bike pump! The lung is then put into a water tank, and the bodek looks for air bubbles. If the lung is still intact, it is kosher. In the United States, a glatt kosher animal's lungs generally have fewer than two adhesions, which permits the task to be done carefully in the limited time available in large plants. Some groups—particularly Jews who originated from countries under Muslim rule during the "dark ages" (i.e., Sephardim) require a total absence of adhesions even in adult animals. Such meat is referred to as *Beit Yosef* meat. Note that young red meat animals must always be without adhesions.

The use of the word *glatt* for any other kosher product, including poultry, is only meant to convey the message that a higher standard is being used. It would be more accurate to use the word *Menhadrin*, meaning a stricter standard, and this word is used on some U.S. products and in other countries. Non-glatt meat and non-menhadrin poultry products encompass a larger percentage of the Kosher marketplace (by volume).

Meat and poultry must be further prepared by properly removing certain veins, arteries, prohibited fats, blood, and the sciatic nerve. This process is called *nikkur* in Hebrew and *treiboring* in Yiddish. The person who is specifically trained to do this is called a *Menacker*. In practical terms this means that only the front quarter cuts of kosher red meat are used in the United States and most Western countries.

To further remove the prohibited blood, red meat and poultry must then be soaked and salted (*melicha*) within 72 hours of slaughter. If this is not possible, then non-glatt meat is specially washed (*begissing*), and this wash procedure may be repeated up to two more times, each time within 72 hours of the previous washing. The soaking is done for a half hour in cool water; thereafter, the salting is done for one hour with all surfaces, including cut surfaces and the inside cavity of a chicken, being covered with ample amounts of salt. The salted meat is then rinsed three times. The salted meat must be able to drain throughout, and all the blood being removed must flow away freely.

The animal's heart must be cut open and the congealed blood removed before beginning the overall soaking and salting process. Once the meat is properly koshered, any remaining "red-liquid" is no longer considered "blood" according to halacha, and the meat can be used without further concern for these issues.

The salt used for koshering must be of a crystal size that is large enough that the crystals will not dissolve within the hour and must be small enough to permit complete coverage of the meat. The salt industry refers to this size crystal as kosher salt. The specific process of salting and soaking meat to make it ready for use is also referred to as *koshering* meat.

Because of its high blood content, liver cannot be soaked and salted, but must instead be broiled to at least over half cooked using special equipment reserved for this purpose. The liver is then rinsed, after which the liver can be used in any way the user wishes. A small amount of salt is sprinkled on the liver. In theory any meat can be broiled instead of soaking and salting. However, this has not been done for so many years that some rabbis no longer accept this alternative.

Some concern has been raised about the salt level in kosher meat. Note that only the surfaces are salted, generally using primal cuts (i.e., 20- 40-pound pieces of meat) and that the penetration of the salt is less than a half centimeter in red meat (New York Department of Agriculture and Markets, personal communication). Many pieces of meat, as consumed, have therefore not been directly subjected to the salt treatment. If salt content in a diet is a very important consideration, then one should cut off all surfaces and not use any of the drippings that come out during cooking. However, much of the salt that goes into the meat at the surface is lost during the cooking process.

Any ingredients or materials that might be derived from animal sources are generally prohibited because of the difficulty of obtaining them from kosher animals. This includes many products that might be used in foods and dietary supplements, such as emulsifiers, stabilizers, and surfactants, particularly those materials that are fat-derived. Very careful rabbinical supervision would be necessary to assure that no animal-derived ingredients are included in kosher food products. Almost all such materials are available in a kosher form derived from plant oils. A possible exception might be a normative mainstream gelatin, which is now being produced from glatt kosher beef hides (see the section on gelatin). Also some rennet, the cheese-coagulating enzyme, is obtained from the dried fourth stomach of a kosher slaughtered milk-fed calf.

There are a few concepts in Jewish law that permit materials to alter their status. The first is *Dvar Hadash*, or new entity. If something undergoes a sufficient transformation, as defined rabbinically, it may become a new entity. Another concept that may help create flexibility for food manufacturers is the concept of "dry as wood," where the drying is defined as natural drying for over a year. The concept is used in part to justify the use of natural calf rennet discussed above: the extraction of a chemical from such a material permits its use when it would not otherwise be permitted. Finally there is the concept of "not fit for either a person or, less critically, for a dog." If a material is unacceptable and would not even be eaten by a dog, then the source is not considered a food, which means that anything derived from it could be kosher. Note, however, that some rabbis argue that if an identifiable object, for example, a bone, is placed into such a mixture and is then recovered, that the item was not necessarily ever unfit for a dog.

### Prohibition of Mixing of Milk and Meat

> Thou shalt not seeth the kid in its mother's milk. (Exodus XXIII:19, Exodus XXXIV:26, Deuteronomy XIV:21)

This passage appears three times in the Torah and is therefore considered a very serious admonition. As a result, the law cannot be violated even for nonfood uses such as pet food. Neither can one derive benefit from such a mixture; therefore, one cannot own a cheeseburger business. The meat side of the equation has been rabbinically extended to include poultry (not fish) as both meat and poultry need to be inspected, deveined, salted, and soaked. The dairy side includes all milk derivatives.

Keeping meat and milk separate in accordance with kosher law requires that the processing and handling of all materials and products fall into one of three categories:

1. A meat product
2. A dairy product

3. A neutral product called *pareve*, *parve*, or *parev*. (For words that are transliterations of Hebrew, such as pareve, multiple English spellings are acceptable.)

The pareve category includes all products that are not classified religiously as meat or dairy. Secular classifications may be defined differently. All plant products are pareve, along with eggs, fish, honey, and lac resin (shellac). These pareve foods can be used with either meat products or dairy products. However, if they are mixed with meat or dairy, they take on the identity of the product they are mixed with; that is, an egg in a cheese soufflé becomes dairy.

A special set of rules applies to fish. Fish can be eaten at the same meal at which meat is eaten, but it cannot be mixed directly with the meat. The dishes used with the fish are generally kept separate and rinsed before they are used with meat or vice versa.

To assure the complete separation of milk and meat, all equipment, utensils, pipes, steam, etc. must be of the properly designated category. If plant materials, like fruit juices, are run through a dairy plant, they would be considered a dairy product under kosher law. Some kosher supervision agencies would permit such a product to be listed as dairy equipment (D.E.) rather than dairy. The D.E. tells the consumer that it does not contain any intentionally added dairy ingredients, but that it was made on dairy equipment. (See the section on kosher and allergies.) If a product with no meat ingredients is made in a meat plant, like a vegetarian vegetable soup, it may be marked "meat equipment (M.E.)." Although one may need to wash the dishes before and after use, the D.E. food can be eaten on meat dishes and the M.E. food on dairy dishes. A significant wait is normally required to use a product with dairy ingredients after one has eaten meat. This can range from three to six hours, depending on the customs (*minhag*) of the area from which the husband of each family came. With the D.E. listing, the consumer can use the D.E. product immediately before or after a meat meal but not *with* a meat meal. Following dairy, the wait before eating meat is much less, usually from a "rinse of the mouth" with water to one hour. Certain dairy foods do require the full wait of three to six hours; when a hard cheese is eaten, the wait is the same as that for meat to dairy. A hard cheese is defined as a cheese that has been aged for over six months or one that is particularly dry and hard like many of the Italian cheeses. Thus, most companies producing cheese for the kosher market usually age their cheese for less than six months, although with proper package marking this is not a religious requirement.

If one wants to make an ingredient or product truly pareve, the plant equipment must undergo a process of equipment kosherization (see the section on equipment kosherization). From a marketing stand point, a pareve designation is most desirable since it has the most uses, both for the kosher and for the non-kosher consumer.

## Kosher: Special Foods

### Grape Products

To be kosher, all grape juice–based products can only be handled by sabbath-observing Jews from grape-pressing to final processing. If the juice is pasteurized (heated or *mevushal* in Hebrew), then it can be handled by any worker as an ordinary kosher ingredient.

### Jewish Cheese (*Gevinas Yisroel*)

Similar to the laws concerning kosher wine production, most kosher supervision organizations require that the supervising rabbi add the coagulating agent, that is, the agent that makes the cheese form a curd, into the vat to ensure that the cheese is kosher. Any cheese that does not meet this requirement is unacceptable.

Kosher whey can be created more easily. If all the ingredients and equipment used during cheese making are kosher, the whey will be kosher as long as the curds and whey have not been heated above 120°F (49°C) before the whey is drained off. This is true even if a rabbi has not added the coagulant. The necessity for Jewish participation in cheese making is that the cheese is a product "fit for a king." Clearly, whey does not fit into this category. There is much more kosher whey available in the United States than kosher cheese.

Increasingly, the dairy industry is seeking to sell more whey to other food companies. Since many of these companies are kosher, there has been growing interest in assuring the kosher status of whey. For example, several manufacturers of Swiss cheese, which has one of the most desirable, whitest wheys, have reduced the temperature at which they work the curds under the whey. Instead of using the traditional 125–127°F (52–53°C), they are using a temperature under 120°F (49°C) to work the curds and to obtain a kosher whey.

But there are challenges to be overcome. Much of the whey is produced in spray driers, which are among the most difficult pieces of equipment to kosherize. The process of cleaning out the entire system is quite time consuming.

Another problem deals with whey cream. Any cream that is separated from cheese at above 120°F (49°C) is subject to the restrictions that come with cheese and is generally not considered kosher. This cream has recently been used to produce butter, which is therefore not considered kosher. Most rabbis had traditionally accepted butter as kosher without supervision as is still the case with milk. The transition to requiring kosher supervision of butter has been difficult. A more detailed article on this, and closely related kosher dairy issues, has been published (Regenstein and Regenstein, 2002a, b, c).

### Cholev Yisroel

Some kosher-observant Jews are concerned about possible adulteration of milk with the milk of non-kosher animals, such as mare's milk or camel's milk, and therefore require that the milk be watched from the time of milking. This Cholev Yisroel milk is required by some of the stricter kosher supervision agencies for all dairy ingredients. Rabbis who accept non-Cholev Yisroel milk in the United States do so for two reasons. First, they believe that the laws in the United States and many other countries are strong enough to assure that adulteration does not occur. Second, the non-kosher milks are worth more money than kosher milks, so there is no incentive to add non-kosher milk to the milk of kosher species.

Farms producing Cholev Yisroel milk would have a Sabbath-observing Jew on the farm whenever milking is taking place, including the Sabbath. The milk tanks on the farm and the tank truck taking the milk to market would both be sealed by the on-site religious supervisor, and then the seal would be broken by the receiving religious supervisor at the milk plant.

### Yashon and Chodesh Flour

On the second day of Passover, Jews traditionally brought a grain offering to the temple in Jerusalem. This served to bless all of the flour that was growing or had already been harvested on that day. Such flour has attained the status of *yashon* (old) flour. All wheat for flour that has not started to grow by the second day of Passover is considered *chodesh* (new) and should not be used until the next Passover. All winter wheat from the Northern Hemisphere is automatically considered yashon. It is more difficult to assure the yashon status of spring wheat, which generally is harvested in August. Manufacturers may receive inquiries from consumers about the source and timing of their wheat and other grain purchases, particularly between August and the next Passover.

### Early Fruit

Another kosher law concerning plants is the requirement that tree fruits not be harvested for benefit until the fourth year. This has been particularly problematic with respect to papaya, a tree fruit that is often grown commercially for less than four years! Discussion and disagreement remains at this time.

### Passover

The Passover holiday occurs in spring and requires observant Jews to avoid eating the usual products made from five prohibited grains: wheat, rye, oats, barley, and spelt (Hebrew: *chometz*). Those observing kosher laws can eat only the specially supervised unleavened bread from wheat (Hebrew: *matzos*) that were prepared especially for the holiday. Once again, some matzos, that is, *schmura* matzos are made to a stricter standard with rabbinical inspection beginning in the field. For other Passover matzo the supervision does not start until the wheat is about to be milled into flour. Matzo made from oats and spelt are now available for consumers with allergies.

Special care is taken to assure that the matzo does not have any time or opportunity to "rise." In some cases this literally means that products are made in cycles of less than 18 minutes. This is likely to be the case for handmade schmura matzo. In continuous large-scale operations, the equipment is constantly vibrating so that there is no opportunity for the dough to rise.

Why 18 minutes? Note that the word for *life* is the two letter Hebrew word *Chai*. Since the Hebrew alphabet is "mapped" to numbers (e.g., Aleph = 1, Bet = 2), the word *Chai* equals the number 18! Thus fermentation, "life," is considered to require 18 minutes to occur. Anything made in less than 18 minutes has not fermented and has, therefore, not violated the prohibitions of Passover. Also, the drinking toast among Jews is *L'Chaim*, to life.

In the middle ages, the rabbis of Europe also made products derived from corn, rice, legumes, mustard seed, buckwheat, and some other plants (Hebrew: *kitnyos*) prohibited for Passover. In addition to the actual flours of these materials, many contemporary rabbis also prohibit derivatives such as corn syrup, cornstarch, and cornstarch derivatives such as citric acid. A small number of rabbis permit the oil from kitnyos materials or liquid kitnyos products and their derivatives, such as corn syrup. The major source of sweeteners and starches used for production of sweet Passover items are either real sugar or potato-derived products such as potato syrup.

Rabbis are concerned with other foodstuffs that are being raised in areas where wheat and other Passover grains are grown. Because of possible cross-contamination, some crops such as fennel and fenugreek are also prohibited for Passover.

During the dark ages, the Jewish communities within Christian countries did not have regular contact with Jews living in Muslim countries. The laws governing these two communities began to drift apart. As a result, today's European, or Ashkenazic, Jewish community has some significantly different laws and customs from the Sephardic Jewish community, which included Spain, North Africa, and the Middle East. Sephardic custom, which is the default in Israel, includes among other rules, no ban on all or some of the kitnyos materials like rice, a beit yosef meat standard of absolutely no lung adhesions on animals, and a willingness to use hindquarter that has been correctly subject to nikkur or *deveining*. With a few exceptions, however, Passover foods in the United States are processed to Ashkenazic standards.

Consumers who regularly use products such as dietary supplements and non-life-threatening drugs will be concerned about obtaining a version of their favorite and/or required product that is acceptable at Passover. For drugs, the prohibition of chometz is of special concern since many Jews do not want any manner of chometz in their home, including drugs, pet feeds, and nonfood items such as rubbing alcohol.

A violation of the laws of Passover is considered biblical grounds for being "separated from the community." This is generally the highest level of prohibition and has led to extra strictness with respect to Passover.

The most stringent kosher consumers only eat whole, unbroken matzos on the first seven days of Passover, the seven days observed by Jews everywhere including Israel. Thus, any prepared food for those seven days (the biblically commanded time) may need to be made without the use of any matzo meal or matzo flour, that is, no *gebruckts* (no broken matzos). However, on the eighth day—which is a rabbinical extension of Passover outside of the land of Israel—these people will also eat products made with less than whole matzos, including the traditional Jewish matzo ball soup.

## Kosher: Other Processing Issues

### Equipment Kosherization

There are three ways to make equipment kosher or to change its status back to pareve from dairy or meat. Rabbis generally frown on going from meat to dairy or vice versa. Most conversions are from dairy to pareve or from treife to one of the categories of kosher. There are a range of process procedures to be considered, depending on the equipment's prior production history.

After a plant or a processing line has been used to produce kosher pareve products, it can be switched to either kosher dairy or kosher meat without a special equipment kosherization step. It can also subsequently be used for halal production (from pareve or dairy lines, not always from meat lines), and then, finally, for non-kosher products. In many cases, a *mashgiach*, the rabbinically approved kosher supervisor, is needed on site for equipment kosherization, so it normally is beneficial to minimize the number of changeovers from one status to another.

The simplest equipment kosherization occurs with equipment that has only been handled cold. This requires a good liquid caustic/soap cleaning, the type of cleaning done normally in most food plants. Some plants do not normally do a wet clean up between runs (e.g., a dry powder packing plant or a chocolate line), and these would need to seek specific rabbinical guidance for the changeover. Materials such as ceramics, rubber, earthenware, and porcelain cannot be koshered because they are considered not "capable" of releasing the flavors trapped within them during the equipment kosherization process. If these materials are found in a processing plant, new materials may be required for production.

Most food-processing equipment is operated at cooking temperatures, generally above 120°F (49°C), the temperature that is rabbinically defined as cooking. However, the exact temperature for cooking depends on the individual rabbi, in that it is the temperature at which he must immediately remove his hand when he puts it into hot water. Recently, through an agreement by the major four mainstream American kosher certifying agencies, most normative kosher supervision agencies in the United States have settled on 120°F (49°C) as the temperature at which foods are cooked, and this figure is used throughout this chapter. (See Dealing with Kosher and Halal Supervision Agencies.)

Equipment that has been used with cooked product must be thoroughly cleaned with liquid caustic/soap before being kosherized. The equipment must then be left idle for 24 hours, after which it is flooded with boiling water, being defined as water between 190°F (88°C) and 212°F (100°C), in the presence of a kosher supervisor. The details depend on the equipment being kosherized. In some cases, particularly foodservice establishments, a *pogem* (bittering agent, oftentimes ammonia) is used in the boiling water in lieu of the 24-hour wait. The absolutely clean equipment, for example, silverware, is put into the ammonia containing boiling water to pick up a bad flavor. This bad flavor is removed by a second boiling with clean water. The 24-hour wait accomplishes the same thing as the ammonia; it turns any good flavors attached to the equipment into bad flavors.

The principles concerning koshering by *hagalah* (boiling water) or *irui* (boiling water poured over a surface) are based on an ancient understanding of the movement of *taam* (flavor) in and out of solid materials. The concepts of taam and its movement between products are also used to analyze the many possible combinations of kosher meat, kosher dairy, and/or non-kosher products interacting accidentally, that is, for analysis after the fact (*b'de-eved*). For real accidents, the rabbis are able to be more lenient than they might be for things that are done intentionally (*l'chatchilla*, i.e., planned ahead of time). In modern times, where kosher supervision in the United States is active (i.e., the rabbis are operating with a contractual agreement and ongoing inspections), there is less room to work with some of these leniencies. In Europe, where rabbis sometimes only make informal visits to plants and report on their visits to their congregants and the greater Jewish community, the rules with respect to after-the-fact issues are sometimes used more freely since the rabbi cannot control, nor is he responsible for, any changes the processing plant may make once he has left the plant.

In the case of ovens or other equipment that use fire, or dry heat, kosherization involves heating the metal until it glows. Again, the supervising rabbi is generally present while this process is taking place. In the case of ovens, particularly large commercial ovens, issues related to odor/vapors and steam must also be considered. Sometimes the same oven can be used sequentially for alternating pareve and dairy baking. The details are beyond the scope of this chapter and require a sophisticated rabbinical analysis to determine which ovens can be used for more than one status without requiring kosherization.

The procedures that must be followed for equipment kosherization, especially for hot equipment, can be quite extensive and time-consuming, so the fewer status conversions, the better. Careful formulating of products and good production planning can minimize the inconvenience. If a conversion is needed, it is often scheduled for early Monday morning, before the production week starts.

### Jewish Cooking and Jewish Baking

In cases where it is necessary for rabbis to "do" the cooking (*Bishul Yisroel*), their contribution must remain independent of the company's activities. Often this means turning on the pilot light. As long as the pilot light remains lit, the rabbi does not have to be present; if it goes out, he must return. With electrical equipment and appliances, it is possible to keep electricity on all the time, using the lowest setting when actual heating is not taking place. The most difficult situation for kosher operations is a gas stove with an electrical starter. Care in selecting equipment can prevent a number of problems.

Baking generally requires Jewish participation, *Pas Yisroel*; that is, the Jew must start the ovens. In addition, if the owner of the bakery is Jewish, there may be a requirement for taking *challah*, a portion of the dough that is removed and needs to be specially handled. Again, the details need to be worked out with the supervising rabbi.

Note that a company that is over 50 percent Jewish management or Jewish ownership is subject to stricter rules, e.g., the taking of challah and the need to observe the Sabbath and other Jewish holidays. To be subject to the less strict rules, some owners sell their business to a gentile for the period of concern, even a single day each week. This is a legally binding contract and, in theory, the gentile owner can renege on his or her informal agreement to legally sell it back at the end of *shabbos* or the end of the holiday. On Passover, the need to do this can be more critical: Any chometz in the possession of a Jew during Passover is forever prohibited in a kosher home. For instance, if a Jewish grocery store receives a shipment of bread during Passover, that bread, even if marked as kosher, though obviously non-Passover, can never be used by an observant kosher-observing Jew.

### Toveling (Immersing Equipment Purchased from a Gentile)

When a Jewish company purchases or takes new or used equipment from gentiles, the equipment must be bathed in a ritual bath (*mikvah*) prior to being equipment kosherized. Equipment from metal and glass requires a blessing; complex items that contain glass or metal may need to be toveled, but may not need a blessing. A mashgiach needs to be present for this activity. A natural body of water can be used instead of the indoor mikvah, especially with large equipment.

### Tithing and Other Israeli Agricultural Laws

In ancient times, products from Israel were subject to special rules concerning tithing for the priests, their helpers, the poor, etc. These are complex laws that only affect products from Israel. There is a rabbinical process for doing the tithing that does not require some of the actual product to be removed from the lot. The land of Israel is also subject to the Sabbath (sabbatical) years, that is, crops from certain years cannot be used. These additional requirements challenge kosher consumers in the United States who arc interested in purchasing and trying Israeli products. Rabbis in Israel arrange for companies to tithe when the products are destined for sale in Israel, but rarely for exports. In 2002, at least one major U.S. kosher supervision agency has begun to arrange for tithing before the product is offered to the consumer in the United States. The details of this process are beyond the scope of this chapter.

## Kosher and Allergies

Many consumers use the kosher markings as a guideline to determine whether food products might meet their special needs, including allergies. There are, however, limitations that the particularly sensitive allergic consumer needs to keep in mind.

1. When equipment is kosherized, or converted from one status to another, the procedure may not yield 100 percent removal of previous materials run on the equipment. This became an issue some years ago when rabbis discovered that the special procedures being used to convert a dairy chocolate line to a pareve chocolate line led to enough dairy contamination that consumers who were very sensitive to dairy allergens were having problems. These lines are koshered without water: either a hot oil or pareve chocolate is run through the line in a quantity sufficient to remove any dairy residual as calculated by the supervising rabbi.

Both Islam and Judaism do not permit practices that will endanger life to occur. As a result, rabbis decided that none of the current religiously acceptable methods for equipment kosherization of chocolate are effective enough to move between dairy and pareve production; therefore, mainstream kosher supervision agencies no longer permit this conversion.

2. Kosher law does permit certain ex-post-facto (after the fact) errors to be negated. Trace amounts of materials accidentally added to a food can be nullified if the amount of offending material is less than 1/60 by *volume* under very specific conditions (i.e., truly added by accident). However, some items can never be negated, such as strong flavor compounds that make a significant impact on the product even at less than 1/60. In deference to their industrial client company's desire to minimize negative publicity, many kosher supervision agencies do *not* announce when they have used this procedure to make a product acceptable. When there is a concern about allergic reactions, however, many rabbis are more willing to alert the public as soon as possible for health and safety reasons.

Products that might be made in a dairy plant—e.g., pareve substitutes for dairy products and some other liquids such as teas and fruit juices—may be produced in plants that have been kosherized, but may not meet a very critical allergy standard. Care in consuming such products is recommended.

3. For labels that say Dairy and Meat Equipment, there are no intentionally added dairy or meat ingredients, but the product is produced on a dairy or meat line without any equipment kosherization. The product is considered pareve with some use restrictions in a kosher home. Again, the more sensitive the allergy, the more caution is advised.

4. In a few instances where pareve or dairy products contain small amounts of fish, such as anchovies in Worcestershire sauce, this ingredient *may* be marked as part of the kosher supervision symbol. Many certifications do not specifically mark this if the fish in the initial material is less than 1/60. Someone who is allergic should always read the ingredient label.

5. At Passover, there is some dispute about derivatives of kitnyos materials, the non-grain materials that are also prohibited for Ashkenazic Jews. A few rabbis permit items such as corn syrup, soybean oil, peanut oil, and similarly derived materials from these extensions. The "proteinaeous" part of these materials is generally not used. Consumers with allergies to these items can therefore purchase these special Passover products from supervision agencies that do *not* permit kitnyos derivatives. With respect to equipment kosherization, supervising rabbis tend to be very strict about the clean-up of the prohibited grains (wheat, rye, oats, barley, and spelt), so these Passover products come closest to meeting potential allergy concerns. This may not be the case with respect to the extended kitnyos prohibitions.

Consumers should not assume that kosher markings ensure the absence of trace amounts of the ingredient to which they are allergic. It is a useful first screen, but products should be carefully tested before assuming everything is acceptable; that is, the allergic person should eat a small portion of the product, and increase the amount consumed slowly, over time, to assure no adverse reaction. People with allergies should get into the habit of checking lot numbers on products and purchasing stable goods with a single lot number in sufficient quantity to meet anticipated needs within the shelf-life expectations of the goods.

How thoroughly are dairy ingredients kept out of a pareve line? The current standard for kosher may not meet the needs of allergic consumers since the dairy powder dust in the air may be sufficient to cause allergy problems. A company might choose to use a special marking on kosher pareve chocolates produced in plants that also produce dairy products to indicate that these are religiously pareve, but may not be sufficiently devoid of dairy allergens for very allergic consumers. Furthermore, they may also want to consider checking the chocolate using one of the modern antibody or similar types of tests.

## Halal

### Halal Dietary Laws

The halal dietary laws define food products as halal (permitted), *haram* (prohibited), and a few items go into the category of *makrooh* (questionable to detestable). The law deals with the following five issues; all but the last are in the animal kingdom.

1. Prohibited animals
2. Prohibition of blood
3. Method of slaughtering/blessing
4. Prohibition of carrion
5. Prohibition of intoxicants

The Islamic dietary laws are derived from the Quran, a revealed book; the Hadith, the traditions of Prophet Muhammad; and through extrapolation of and deduction from the Quran and the Hadith by Muslim jurists.

Approximately 90 percent of Muslims are Sunni, while the other 10 percent are Shiia. This chapter will general describe Sunni practice.

There are 11 generally accepted principles pertaining to halal and haram in Islam for providing guidance to Muslims in their customary practices.

1. The basic principle is that all things created by Allah are permitted, with a few exceptions that are prohibited. Those exceptions include pork, blood, meat of animals that died of causes other than proper slaughtering, food that has been dedicated or immolated to someone other than Allah, alcohol, intoxicants, and inappropriately used drugs.
2. To make lawful and unlawful is the right of Allah alone. No human being, no matter how pious or powerful, may take it into his hands to change it.
3. Prohibiting what is permitted and permitting what is prohibited is similar to ascribing partners to Allah. This is a sin of the highest degree that makes one fall out of the sphere of Islam.
4. The basic reasons for the prohibition of things are due to impurity and harmfulness. A Muslim is not supposed to question exactly why or how something is unclean or harmful in what Allah has prohibited. There might be obvious reasons and there might be obscure reasons.
5. What is permitted is sufficient, and what is prohibited is then superfluous. Allah prohibited only things that are unnecessary or dispensable while providing better alternatives. People can survive and live better without consuming unhealthful carrion, unhealthful pork, unhealthful blood, and the root of many vices, alcohol.
6. Whatever is conducive to the prohibited is in itself prohibited. If something is prohibited, anything leading to it is also prohibited.
7. Falsely representing unlawful as lawful is prohibited. It is unlawful to make flimsy excuses, to consume something that is prohibited, such as drinking alcohol for supposedly medical reasons.
8. Good intentions do not make the unlawful acceptable. Whenever any permissible action of the believer is accompanied by a good intention, his action becomes an act of worship. In the case of haram, it remains haram, no matter how good the intention or how honorable the purpose may be. Islam does not endorse employing a haram means to achieve a praiseworthy end. The religion indeed insists not only that the goal be honorable, but also that the means chosen to achieve it be lawful and proper. Islamic laws demand that the right should be secured solely through just means.

9. Doubtful things should be avoided. There is a gray area between clearly lawful and clearly unlawful. This is the area of "what is doubtful." Islam considers it an act of piety for Muslims to avoid doubtful things, to stay clear of the unlawful.
10. Unlawful things are prohibited to everyone alike. Islamic laws are universally applicable to all races, creeds, and sexes. There is no favored treatment of a privileged class. Actually, in Islam, there are no privileged classes; hence, the question of preferential treatment does not arise. This principle applies not only among Muslims, but between Muslims and non-Muslims as well.
11. Necessity dictates exceptions. The range of prohibited things in Islam is quite limited, but emphasis on observing the prohibitions is very strong. At the same time, Islam is not oblivious to the exigencies of life, to their magnitude, or to human weakness and capacity to face them. A Muslim is permitted, under the compulsion of necessity, to eat a prohibited food to ensure survival, but only in quantities sufficient to remove the necessity and avoid starvation.

## Prohibited and Permitted Animals

Meat of pigs, boars, and swine is strictly prohibited, as are the carnivorous animals such as lions, tigers, cheetahs, cats, dogs, and wolves. Also prohibited are birds of prey such as eagles, falcons, osprey, kites, and vultures.

Meat of domesticated animals such as ruminants with split hooves (e.g., cattle, sheep, goat, lamb) is allowed for food, as are camels and buffaloes. Also permitted are the birds that do not use their claws to hold down food, such as chickens, turkeys, ducks, geese, pigeons, doves, partridges, quails, sparrows, emus, and ostriches. Some of the animals and birds are permitted only under special circumstances or with certain conditions. Horse meat may be permitted for consumption under some distressing conditions, discussion of which is beyond the scope of this chapter. The animals fed unclean or filthy feed, for example, formulated with biosolids (sewage) or protein from tankage, must be quarantined and placed on clean feed for a period varying from 3 to 40 days before slaughter to cleanse their systems.

Food from the sea, namely, fish and seafood, are the most controversial among various denominations of Muslims. Certain groups, particularly Shiia, only accept fish with scales as halal, while others consider as halal everything that lives in the water all the time. Consequently, prawns, lobsters, crabs, and clams are halal but may be detested (Makrooh) by some and, hence, not consumed. Animals that live both in water and on land (amphibians) such as frogs, turtles, crocodiles, and seals are also not consumed by the majority of observant Muslims.

There is no clear status of insects established in Islam, except that locust is specifically mentioned as halal. Insects are generally considered neutral. However, from deduction of the laws, it seems that both helpful insects like bees, ants, and spiders, and harmful or dirty creatures like lice, flies, and mosquitoes are all prohibited as food. Among the by-products from insects, use of honey was very highly recommended by Prophet Muhammad. Other products such as royal jelly, wax, shellac, and carmine are acceptable to be used without restrictions by most; however, some may consider shellac and carmine Makrooh or offensive to their psyche.

Eggs and milk from permitted animals are also permitted for Muslim consumption. Milk from cows, goats, sheep, and buffaloes is halal. Unlike kosher, there is no restriction on mixing meat and milk.

### Prohibition of Blood

According to the Quranic verses, blood that pours forth is prohibited for consumption. It includes blood of permitted and nonpermitted animals alike. Liquid blood is generally not offered for sale or consumed by Muslims or non-Muslims, but products made with and from blood are available. There is general agreement among Muslim scholars that anything made from blood is unacceptable. Products such as blood sausage and ingredients such as blood albumin are either haram or questionable at best and should be avoided for product formulations.

### Proper Slaughtering of Permitted Animals

There are special requirements for slaughtering the animal.

- An animal must be of a halal species.
- It must be slaughtered by an adult and sane (i.e., mentally competent) Muslim.
- G-d must be invoked by name at the time of slaughter.
- Slaughter must be done by cutting the throat in a manner that induces rapid and complete bleeding, resulting in the quickest death. The generally accepted method is to cut at least three of the four passages: carotids, jugulars, trachea, and esophagus. Some Islamic scholars do accept machine slaughter, particularly of poultry. In recent years,

however, the trend has gone back toward requiring hand slaughter of these animals.

The meat of animals thus slaughtered is called *zabiha* or *dhabiha* meat. "Verily Allah has prescribed proficiency in all things. Thus, if you kill, kill well; and if you perform dhabiha, perform it well. Let each one of you sharpen his blade and let him spare suffering to the animal he slays" (Khan 1991).

Islam places great emphasis on gentle and humane treatment of animals, especially before and during slaughter. Some of the conditions include giving the animal proper rest and water, avoiding conditions that create stress, not sharpening the knife in front of the animals, using a very sharp knife to slit the throat, etc. Only after the blood is allowed to drain completely from the animal and the animal has become lifeless can the dismemberment, cutting off horns, ears, legs, etc. commence. Unlike kosher, soaking and salting of the carcass is not required for halal; halal meat, once slaughtered, is therefore treated like other commercial meat. Animal-derived food ingredients like emulsifiers, tallow, and enzymes must be made from animals slaughtered by a Muslim to be halal.

Hunting of permitted wild animals (like deer) and birds (like doves, pheasants, and quail) is permitted for the purpose of eating but not merely for deriving pleasure out of killing an animal. Hunting during the pilgrimage to Makkah (Mecca) and within the defined boundaries of the holy city of Makkah is strictly prohibited. Hunting is permitted with any tools, for instance, guns, arrows, spears, or trapping. Trained dogs may also be used for catching or retrieving the hunt. The name of Allah may be pronounced at the time of releasing the tool rather than catching of the hunt. The animal has to be bled by slitting the throat as soon as the animal is caught. If the blessing is made at the time of pulling the trigger or shooting an arrow and the hunted animal dies before the hunter reaches it, it would still be halal as long as slaughter is performed and some blood comes out. Fish and seafood may be hunted or caught by any reasonable means available as long as it is done humanely, and no blessing needs to be said.

The requirements of proper slaughtering and bleeding are applicable to land animals and birds. Fish and other creatures that live in water need not be ritually slaughtered. Similarly there is no special method of killing the locust.

The meat of the animals that die of natural causes (e.g., diseases, being gored by other animals, being strangled, falling from a height, through beating, or killed by wild beasts) is unlawful to be eaten, unless one saves such animals by slaughtering before they actually become lifeless. A fish that dies naturally and is floating on water or lying out of water is still halal as long as it does not show any signs of decay or deterioration.

### Meat of Animals Killed by the Ahl-al-Kitab

There has been much discussion and controversy among Muslim consumers as well as Islamic scholars about the permissibility of consuming meat of animals killed by the *Ahl-al-Kitab*, or people of the book, meaning, among certain other faith communities, Jews and Christians. The issue focuses on whether meat prepared in the manner practiced by either faith would be permitted for Muslims.

In the Holy Quran, this issue is presented only once in Sura V, verse 5, in the following words:

> This day all good things are made lawful for you. The food of those who have received the Scripture is lawful for you, and your food is lawful for them.

This verse addresses the Muslims and seems to establish a social context where Muslims, Jews, and Christians could interact with each other. It points toward two sides of the issue: first, "the food of the people of the book is lawful for you," and second, "your food is lawful for them."

In most discussions, scholars try to deal with the first part (food of Ahl-al-Kitab) and ignore the second part (food of Muslims) altogether, leaving that decision to the people of the book.

As far as the first part of the ruling is concerned, Muslims are allowed to eat the food of the Jews and Christians as long as it does not violate the first part of this verse, "this day all good and wholesome things have been made lawful for you" (Quran V:6).

The majority of Islamic scholars are of the opinion that the food of the Ahl-al-Kitab must meet the criteria established for halal and wholesome food including proper slaughter of animals. They believe that the following verse establishes a strict requirement for Muslims.

> And eat not of that whereupon Allah's name hath not been mentioned, for lo! It is abomination. (Quran VI:121)

However, some Islamic scholars are of the opinion that the above verse does not apply to the food of Ahl-al-Kitab, and there is no need to mention the name of G-d at the time of slaughtering (Al-Qaradawi 1984). It is up to the regulatory agencies in the halal food-importing countries, halal certifiers for export or domestic consumption, or the individual Muslim consumers to decide how to interpret these verses.

For the Muslims who want to follow requirements of verse VI:121, meat (red meat and poultry) of the Ahl-al-Kitab may not meet halal standards. In addition, as discussed elsewhere in this chapter, dairy and pareve kosher products may contain alcohol (e.g., in flavors), and some more lenient kosher supervisions as defined above will permit products that contain animal-based ingredients that may also be unacceptable to the halal-observing consumer.

### Prohibition of Alcohol and Intoxicants

Consumption of alcoholic drinks and other intoxicants is prohibited according to the Quran (V:90–91), as follows:

> O you who believe! Fermented drinks and games of chance, and idols and divining arrows are only an infamy of Satan's handiwork. Leave it aside in order that you may prosper. Only would Satan sow hatred and strife among you, by alcohol, and games of chance, and turn you aside from the remembrance of Allah, and from prayer: Will you not, therefore, abstain from them?

The Arabic term used for alcohol in the Quran is *khamr*, which means "that which has been fermented," and implies not only alcoholic beverages such as wine, beer, whiskey, and brandy, but has been taken to imply all things that intoxicate or affect one's thought process. Although there is no allowance for added alcohol in any beverage like soft drinks, small amounts of alcohol contributed from food ingredients may be considered an impurity and hence ignored. Synthetic or grain alcohol may be used in food processing for extraction, precipitation, dissolving, and other reasons, as long as the amount of alcohol remaining in the final product is very small, generally below 0.1 percent. Each importing country may have its own guidelines, which must be understood by the exporters and strictly adhered to.

In the West, food may be cooked in alcohol to enhance the flavor or to impart distinctive flavor notes. Wine is the most common form of alcohol used in cooking. While one may think that all of the added alcohol evaporates or burns off during cooking, the fact is that it does not. The alcohol retained in the food products varies depending upon the cooking method.

Even after cooking for 2.5 hours, up to 5 percent alcohol remains in the food (Larsen, 1995). Although there is little chance of intoxication by eating such food, the use of alcoholic drinks in cooking is categorically prohibited.

#### Halal Cooking, Food Processing, and Sanitation

Alcohol may not be used in cooking. Otherwise, there are no restrictions about cooking in Islam, as long as the kitchen is free from haram foods and ingredients. There is no need to keep two sets of utensils, one for meat and the other for dairy, as in kosher cooking.

In food companies, haram materials should be kept segregated from halal materials. The equipment used for non-halal products must be thoroughly cleansed using proper techniques of acids, bases, detergents, and hot water. As a general rule, kosher cleaning procedures would be adequate for halal, too. If the equipment is used for haram products, it must be properly cleaned, sometimes by using an abrasive material, blessed by a Muslim inspector, and finally be rinsed with hot water seven times.

## Both Kosher and Halal

### Science

#### Gelatin

Important in many food products, gelatin is probably the most controversial of all modern kosher and halal ingredients. Gelatin can be derived from pork skin, beef bones, or beef skin. In recent years, some gelatins from fish skins have also entered the market. Fish gelatins can be produced kosher and halal with proper supervision and are acceptable to almost all of the mainstream religious supervision organizations.

Most currently available gelatins—even if called "kosher"—are not acceptable to the mainstream U.S. kosher supervision organizations and to Islamic scholars. Many are, in fact, totally unacceptable to halal consumers because they may be pork-based gelatin.

A recent development has been the manufacture of kosher gelatin from the hides of kosher-slaughtered and -inspected cattle. It has been available in limited supply at great expense, and this gelatin has been accepted by the mainstream and even some of the stricter kosher supervision agencies. Similarly, at least two major manufacturers are currently producing certified halal gelatin from cattle bones of animals that have been slaughtered by Muslims. Vegetarian capsules are also available, made with starch, cellulose, or other vegetable ingredients.

One finds a wide range of attitudes toward gelatin among the lenient kosher supervision agencies. The most liberal view holds that gelatin, being made from bones and skin, is not being made from a food (flesh). Furthermore, the process used to make the product goes through a stage where the product is so unfit that it is not edible by man or dog and as such becomes a new entity. Rabbis holding this view may accept pork gelatin. Most water gelatin desserts with a generic K on the package follow this ruling.

Other rabbis permit only gelatin from beef bones and hides, and not pork. Still other rabbis accept only "India dry bones" as a source of beef gelatin. These bones, found naturally in India from the animals that fell and died in the fields (because of the Hindu custom of not killing the cows), are aged for over a year and are "dry as wood"; additional religious laws exist for permitting these materials. Again, *none* of these products is accepted by the mainstream kosher or halal supervisions and are therefore not accepted by a significant part of the kosher and halal community.

## Biotechnology

Rabbis and Islamic scholars currently accept products made by simple genetic engineering; *chymosin* (rennin) was accepted by the rabbis about a half a year before it was accepted by the U.S. Food and Drug Administration. The basis for this decision involves the fact that the gene isolated from a non-kosher source is far below "visible." Subsequently, it is copied many times in vitro and then eventually injected into a host that is then reproduced many times. Thus, the original source of the gene is essentially totally lost by the time the food product appears. The production conditions in the fermenters must still be kosher or halal, that is, the ingredients and the fermenter, and any subsequent processing, must use kosher or halal equipment and ingredients of the appropriate status. A product produced in a dairy medium, for example, extracted from cow's milk, would be dairy. Mainstream rabbis may approve porcine lipase made through biotechnology when it becomes available, if all the other conditions are kosher. Islamic scholars are still considering the issue of products with a porcine gene; although a final ruling has not been established, the leaning seems to be toward rejecting such materials. If the gene for a porcine product were synthesized, that is, it did *not* come directly from the pig, Islamic scholars still need to determine if they will accept it. Because the religious leaders of both communities have not yet determined the status of more complex genetic manipulations; such a discussion is premature.

### Pet Food

Jews who observe the kosher laws can feed their domestic animals pet food that contains pork or other prohibited meats. They cannot feed their animals products that contain a mixture of milk and meat. On Passover their pet food can contain kitnyos, but not chometz. Although pets, even in a halal-observant home, can be fed anything, many individual Muslims prefer to use pet foods without pork and other prohibited materials.

### Health

As described above, the Muslim halal laws are focused on health. Although many people believe that the kosher laws are also considered to be among the laws that were given for people's benefit, this is not the case. One of the few exceptions is the rule concerning the mixing of meat and fish, which was rabbinically instituted to avoid a problem with a particular fish, which when eaten with meat, made people sick. Because this is one of the few laws that is a health law, the Conservative movement recently saw fit to rule that it is no longer valid since we cannot identify the fish nor is there any evidence currently of such a problem.

The most common health-related aspect of the kosher laws that is cited is the prevention of trichinosis in pork. This argument has three weaknesses. First, all flesh products can be a source of pathogens. The full cooking that is traditional in the Jewish community gives better pathogen control, although there seems to be no religious law or custom (minhag) that mandates this practice. Second, the presence of trichinosis in mummified pork has not been demonstrated. And third, the incubation period for trichinosis, 10–14 days, makes it unlikely that the ancient Israelites would have figured out the correlation at that time.

## Regulatory Agencies

### Dealing with Kosher and Halal Supervision Agencies

In practical terms, the food industry works with kosher and halal supervision agencies to obtain permission to use the supervision agency's trademark symbol on their products. In this way, the industry can make claims in the marketplace that are legal and, more importantly, credible to those intentionally purchasing these products. This potential choice provides a significant potential opportunity.

Kosher or halal supervision is taken on by a company to expand its market opportunities. It is a business investment.

What criteria should a company use to select a supervision agency? Supervision fees must be taken into account, and the agency's name recognition is a consideration. Other important considerations include: (1) responsiveness in handling paperwork, in providing mashgiachs or Muslim inspectors at the plants as needed on a timely basis, and in doing routine inspections at a defined frequency during the year, anywhere from twice a year to every day (including continuous) depending on the nature of the production; (2) willingness to work with the company on problem solving; (3) ability to clearly explain their kosher or halal standards and their fee structure. And, of course, one should also consider if the personal chemistry is right for a working relationship, and if their religious standards meet the company's needs in the marketplace.

One of the hardest issues for the food industry to deal with in day-to-day kosher activities is the existence of so many different kosher supervision agencies. Halal has fewer agencies, but still has many standards. How does this impact the food companies? How do the Jewish kosher or Muslim halal consumers perceive these different groups? Because there has not been a central ruling authority for many years in either religion, different rabbis and Muslim inspectors follow different traditions with respect to their dietary standards. Some authorities tend to follow the more lenient standards, while others follow more stringent standards. The trend in the mainstream kosher community today is toward a more stringent standard since some of the previous leniencies were considered undesirable but were tolerated when fewer alternatives were available. The mainstream Islamic scholars also seem to be moving toward tighter standards so that the approved products are acceptable to a larger audience.

One can generally divide the kosher supervision agencies into three broad categories. First there are the large organizations that dominate the supervision of larger food companies: the OU (Union of Orthodox Jewish Congregations, Manhattan, New York); the OK (Organized Kashrus Laboratories, Brooklyn, New York); the Star-K (Baltimore, Maryland); and the Kof-K (Teaneck, New Jersey); all four of which are nationwide and mainstream organizations.

### Normative Mainstream Kosher Supervision

Let's take a quick digression to explain the concept of normative mainstream kosher supervision. The concept of a normative mainstream U.S. kosher standard was the outcome of surveys of kosher foods in the supermarket by a food science class on kosher and halal food regulations taught each year at Cornell University. Over 40 percent of the grocery products in the supermarket have a kosher certification, and almost all of these reflect the same normative U.S. standard. This de facto kosher standard in the United States is represented by the major national supervision agencies, the OU, the OK, the Kof-K, and the Star-K, and recently the Half-Moon K. Many of the smaller kosher supervision agencies also use this same standard. There are numerous trademarked kosher symbols, over 590 at last count (*Kashrus Magazine*, 2003), used around the world that identify the kosher supervision agencies and, indirectly, their different, and sometimes controversial, standards of kosher supervision. Some are more lenient than the normative standard, while others are stricter. The letter K cannot be trademarked; any person or company can put a K on a product for any reason. Symbol look-alikes sometimes occur both as kosher markings and as symbols used for other purposes, for example, the circle-K of a convenience store chain.

In addition to the national supervision agencies, there are smaller private organizations and many local community organizations that provide equivalent religious standards of supervision. As such, products accepted by any of the normative mainstream organizations will, with an occasional exception, be accepted by other similar organizations. The local organizations may have a bigger stake in the local community. They may be more accessible and easier to work with. Although often having less technical expertise, they may be backed up by one of the national organizations. For a company marketing nationally, a limitation may be whether consumers elsewhere in the United States know and recognize the local kosher symbol.

The second category of kosher supervision (more stringent than normative mainstream) includes individual rabbis, generally associated with the Hassidic communities, that is, groups with standards beyond the normative Orthodox standard. There are special food brands that cater specifically to these needs. Many of the products used in these communities require continuous rabbinical supervision rather than the occasional supervision used by the mainstream organizations for production-line products.

For local processing (e.g., bakery, deli, restaurant, butcher shop, etc.) either continuous or fairly regular supervision is the norm—often with a local rabbi visiting almost every day.

The third level is composed mainly of individual rabbis who are more lenient than the mainstream standard. Many of these rabbis are Orthodox; some may be Conservative. Their standards are based on their interpretation of the kosher laws. Employing a more lenient rabbi means that the food processor cuts out more of the mainstream and stricter markets; this is a retail marketing decision that each company makes for itself. More lenient supervisions are sometimes the only ones that will certify a product with a special problem that causes other supervisions to reject it. For example, since fish blocks, which are used for fish sticks and portions, are produced around the world, it is difficult to get proper on-site supervision to assure that all fillets in the block are really the species on the label. As a result, only a lenient rabbi will accept such blocks based on a rule of the majority and the assumption that governmental authorities are also monitoring this situation.

Some companies have used the generic K, that is, the letter K, which cannot be trademarked. The generic K is not trusted by many educated kosher consumers. They realize that the symbol is generally used by one of the more lenient supervisions. A few large, national brands have used the generic K for many years even though they have normative mainstream supervisions. Most kosher consumers are aware of these few companies, for example, PepsiCo, Kellogg Co. Although these companies do not seem to lose market share because of this decision, it is still viewed suspiciously by some consumers.

The Muslim community has only one mainstream agency at this time, the Islamic Food and Nutrition Council of America (IFANCA) in Chicago, Illinois, which is also recognized by many Muslim countries. Other Muslim groups are entering the field, but their standards are not as well defined. Some groups and individuals have resorted to certifying their own products. If one has any interest in exporting to Muslim countries or countries with a significant Muslim population, it is extremely important to know which countries will accept the supervision of which agencies.

In recent years we have started to see products that have dual halal and kosher certification. The first were the military MREs (meals ready-to-eat), but the market has since expanded to other industrial (ingredients) and consumer products. Some of the civilian versions of the MREs are available in long-term shelf-stable (nonrefrigerated) form that makes them convenient for use (Jackson 2000). Meat products are either glatt kosher or dhabiha halal, while the pareve and dairy products have the dual certification.

Ingredient companies should be particularly careful in selecting a supervision agency. They should try to use a mainstream kosher or halal supervision agency because most kosher or halal food-manufacturing companies will require such supervision. The ability to sell to as many customers as possible requires a broadly acceptable standard. Unless an ingredient is acceptable to the mainstream, it is almost impossible to gain the benefit of having a kosher ingredient for sale. Ingredient companies need to pay attention to the status of the kosher product; a pareve product is preferred over a dairy product because it has broader potential use.

Food companies will have to pay increasingly more attention to halal standards. In many cases a few changes make it possible to permit kosher products to also serve the halal community, for example, the true absence of animal products and care to assure that any residual alcohol in products is below 0.1 percent. Again a supervision standard acceptable in all or most Muslim countries is desirable.

Note that the 0.1 percent alcohol in the finished product standard is used by IFANCA and seems to be acceptable to the leadership of most halal communities. However, many halal consumers are not familiar with this standard at this time, so further education will be necessary.

There is some amount of interchangeability between kosher supervision agencies. A system of certification letters is used to provide information from the certifying rabbi concerning the products he has approved. The supervising rabbi certifies that a particular plant produces kosher products or that only products with certain labels or codes are kosher under his supervision. To prevent fraud, it is helpful if these letters are renewed every year and dated with both a starting and ending date. These letters are the mainstay of how food companies and other kosher supervision agencies establish the kosher status of ingredients as ingredients move in commerce. Consumers may also ask to see such letters. Obviously a kosher supervision agency will only accept letters from agencies they find acceptable. That acceptance decision depends on two components: the actual kosher standards of the other agency and an assessment of how well they operate and enforce their supervision.

There are, of course, periodic recalls of specific products for various kosher defects that would prevent their use. *Kashrus Magazine* and its websites www.kashrusmagazine.com and www.kashrut.com try to provide up-to-date listings of products with problems, both of consumer items and industrial ingredients. Such a system of certification letters is also used in the Muslim community.

The kosher or halal symbol of the certifying agency or individual doing the certification may appear on the packaging. In some industrial situations, where kosher and non-kosher (or halal and non-halal) products are similar, some sort of color-coding of product labels and packages may also be used.

Most of these religious supervision symbols are trademarks that are duly registered. In a few cases, multiple rabbis have used the same kosher symbol, causing consumer confusion.

Three additional notes about kosher and halal markings on products.

1. To ensure that labels are marked properly, it is the responsibility of the food company to show its labels to its certifying agency before printing. This responsibility includes both the agency symbol and the documentation establishing its kosher status, for instance, dairy or pareve. It is the responsibility of the kosher supervision agency to review these labels carefully. Many kosher supervision agencies currently do not require that *pareve* be marked on products; others do not use the dairy marking. This causes consumer confusion, which could be avoided if every kosher product had its status marked. In addition to providing the proper information, having each product marked with its status would challenge everyone to pay more attention to properly marking products, avoiding recalls, or announcements of mislabeled products. The letter P/p has been used for both Passover and pareve. We suggest using the letter n for pareve, for *neutral*, consistent with the D for dairy and the M for meat.

2. The labels for private label products with specific agency symbols on their labels should not be moved between plants and cannot be used if supervision changes. This is why some companies, both private label and branded, use the generic K. Thus, if the kosher supervision agency changes, the label can still be used. The sophisticated kosher consumer, however, is more and more uncomfortable with this symbol. A major concern is that the labels may be too easily moved between plants, including plants that are not kosher.

The Kashrus Council of Toronto requires that each label have a plant number on it. This prevents the movement of labels between plants of the same company. They are the only agency that currently requires this additional safeguard.

If a company uses the generic K, the customer service and sales departments of the company—and those people representing the company at trade shows—need to know who the certifying rabbi is.

3. In many Muslim countries a generic halal symbol, the word Halal in Arabic in a circle, has been used indiscriminately. Muslim consumers do not have much faith in such a symbol. In North America some small companies have used similar generic markings or just the word Halal or letter H to signify that food is halal, but such symbols are not widely accepted. The Islamic Food and Nutrition Council of America uses a registered trademark logo of the letter M inside a closed crescent. Another agency, the Muslim Consumer Group, uses a triangle H as their logo. Many other halal logos have started to appear on packages in North America, usually on imported foods. Indonesia, Malaysia, Singapore, and Thailand have central halal control bodies, each with their unique logo. As the volume of halal products offered in local and international markets grows, it is expected that determining the standards for a halal certification will become more complex.

## Federal and State Regulations

Making a claim of kosher on a product was a legal claim in the United States. The *Code of Federal Regulations* (21*CFR*101.29) had a paragraph indicating that such a claim must be appropriate but this was removed a few years ago. Approximately 20 states, some U.S. counties, and some cities have laws specifically regulating the claim of kosher. Many of these laws refer to Orthodox Hebrew practice or some variant of this term, for example, reference to specific Jewish documents, and their legality at this time is subject to further court interpretation.

New York state probably has the most extensive set of state kosher laws. These laws, however, were declared unconstitutional by the Federal District Court for Eastern New York. The verdict was upheld by the Federal Court of Appeals for the Second District. The appeal to the entire Second District for "en banc" review was denied. Just recently (in 2003), the Supreme Court of the United States refused to hear an appeal, so the state of New York is now working on developing a new law that will be constitutional. The original law includes a requirement to register kosher products with the Kosher Enforcement Bureau of the Department of Agriculture and Markets. This part of the law was not declared unconstitutional and is still being enforced, and for now companies should certainly continue to comply.

The state of New Jersey has relatively new kosher laws because the state's original kosher laws were declared unconstitutional by the New Jersey State Supreme Court. It was the same problem as New York, that is, requiring an Orthodox standard. The new laws focus specifically on consumer right-to-know issues and truth in labeling. They avoid having the state of New Jersey define kosher. Rather, the food producer defines its terms and is held to that standard. Rabbis (or anyone else) providing supervision can then declare the information consumers need to know to make an informed decision. New Jersey in July 2000 passed a bill extending the same protection to the Muslim community. Enabling regulations are being prepared, and as of this writing appear to just be coming into play. We hope that a similar legal approach to kosher and halal regulations will be adopted by the other states, particularly New York, and that all of the states with kosher laws will extend the same protection to food products produced with halal certification.

Since the New Jersey law was passed, four other states, Minnesota, Illinois, Michigan, and California have passed halal legislation. The new law in Illinois is of concern because of the potential for a violation of the separation of church and state in the First Amendment to the Constitution, a part of the Bill of Rights: The new law states, "The word 'halal' is here defined to mean a strict compliance with every Islamic law and custom pertaining and relating to the . . . ." We anticipate interesting legal follow-up, especially after the recent Supreme Court rejection of New York State's appeal.

## Animal Welfare

The largest fast food chains in the United States are seeking to develop a set of animal welfare standards that determine the purchasing of products they use in the United States and in many other markets. As it became clear that it was not ideal to have each supermarket chain and each chain restaurant come up with its own standards, the FMI (the trade association for many of the supermarkets in North America) and the NCCR appointed an animal welfare committee to come up with a single national animal welfare standard for each species as well as for animal slaughter and poultry slaughter. It is anticipated that these standards will be predominantly based on the animal welfare guidelines developed by the trade associations of production agriculture and meat processing, but that there will be issues on which the FMI/NCCR standards may diverge. The development of standards will have a major impact on animal agriculture throughout the United States and eventually around the world. These standards generally raise the bar in the United States for animal welfare, but are less aggressive than those currently being applied in Europe. The committee is exploring significant improvements in how *all* animals are raised and slaughtered. Initially the effort has focused on each of the trade associations concerned with the major animals of production agriculture (beef, dairy, chicken, turkey, egg layers, and pigs), and with the slaughter process for these animals.

Animal welfare issues that arise in religious slaughter are incorporated in the FMI/NCCR committee work. A discussion of issues appears in Regenstein and Grandin (2002) along with recommendations for auditable standards that will be used by the FMI/NCCR auditors. These religious slaughter standards are consistent with the American Meat Institute requirements that all religious slaughter be done with the animals in an upright position (for mammals). The standard shackling line is also permitted for poultry religious slaughter. For more information, please see the FMI website at http://www.fmi.org.

## References and Additional Readings

Al-Qaradawi, Y. 1984. *The Lawful and Prohibited in Islam.* Beirut, Lebanon: The Holy Quran Publishing House.

Blech, Zushe. 2004. Royal Jelly. In *Kosher Food Production.* Ames: Iowa State Press. In preparation.

Chaudry, Muhammad M. 1992. Islamic food laws: Philosophical basis and practical implications. *Food Technology* 46(10):92.

Chaudry, Muhammad M. and Joe M. Regenstein. 1994. Implications of biotechnology and genetic engineering for kosher and halal foods. *Trends in Food Science and Technology* 5:165–168.

Chaudry, Muhammad M. and Joe M. Regenstein. 2000. Muslim dietary laws: Food processing and marketing. *Encyclopedia of Food Science:* 1682–1684.

Egan, M. 2002. Overview of halal from Agri-Canada perspective. Presented at the Fourth International Halal Food Conference, April 21–23. Sheraton Gateway Hotel, Toronto, Canada.

Grunfeld, I. 1972. *The Jewish Dietary Laws.* p. 11–12. London: The Soncino Press.

Jackson, Mary Anne. 2000. Getting religion—for your product, that is. *Food Technology* 54(7):60–66.

Khan, G.M. 1991. *Al-Dhabah, Slaying Animals for Food the Islamic Way*. p. 19–20. Jeddah, Saudi Arabia: Abul Qasim Bookstore.

Larsen, J. 1995. Ask the Dietitian. Hopkins, MN: Hopkins Technology, LLC., (http://www.dietitian.com/alcohol.html). Accessed April 24, 2003.

Ratzersdorfer, Micheline, Joe M. Regenstein, and Laura M. Letson. 1988. Appendix 5: Poultry Plant Visits. In *A Shopping Guide for the Kosher Consumer*, edited by Joe M. Regenstein, Carrie E. Regenstein, and Laura M. Letson for Governor Mario Cuomo. pp. 16–24. Albany, N.Y.: State of New York.

Regenstein, Joe M. 1994. Health aspects of kosher foods. *Activities Report and Minutes of Work Groups & Sub-Work Groups of the R & D Associates* 46(1):77–83.

Regenstein, Joe M. 2002. Study room: Halal and Kosher: The Muslim and Jewish Dietary Laws. http://cybertower.cornell.edu.

Regenstein, Joe M., Muhammad M. Chaudry, and Carrie E. Regenstein. 2003. The kosher and halal food laws. *Comprehensive Reviews in Food Science and Food Safety* 2(3):111–127. On-line journal at http://www.ift.org.

Regenstein, Joe M. and Temple Grandin. 2002. Kosher and halal animal welfare standards. *Institute of Food Technologists Religious and Ethnic Foods Division Newsletter* 5(1):3–16.

Regenstein, Joe M. and Carrie E. Regenstein. 1979. An introduction to the kosher (dietary) laws for food scientists and food processors. *Food Technology* 33(1):89–99.

Regenstein, Joe M. and Carrie E. Regenstein. 1988. The kosher dietary laws and their implementation in the food industry. *Food Technology* 42(6):86, 88–94.

Regenstein, Joe M. and Carrie E. Regenstein. 2000. Kosher foods and food processing. *Encyclopdedia of Food Science*: 1449–1453.

Regenstein, Joe M. and Carrie E. Regenstein. 2002a. The story behind kosher dairy products such as kosher cheese and whey cream. *Cheese Reporter* 127(4):8, 16, 20.

Regenstein, Joe M. and Carrie E. Regenstein. 2002b. What kosher cheese entails. *Cheese Marketing News* 22(31):4, 10.

Regenstein, Joe M. and Carrie E. Regenstein. 2002c. Kosher by-products requirements. *Cheese Marketing News* 22(32):4, 12.

Riaz, Mian N. and Muhammad M. Chaudry. 2003. *Halal Food Production*. Boca Raton, FL: CRC Press.

Wikler, Y. 2003. Kosher supervision guide. *Kashrus Magazine* 24(1):30–95.

CHAPTER 10

# Biotechnology

Patricia Curtis, Auburn University
Brooke Caudill

All living organisms are made up of cells containing deoxyribonucleic acid (DNA). The information provided by DNA is what determines the characteristics of living organisms. DNA is usually transferred by traditional breeding methods such as natural selection and crossbreeding. Scientists have discovered that DNA is interchangeable among animals, plants, bacteria, and other organisms. This means that in most cases genes that determine desirable traits can be transferred from one plant or animal to another. Genetic engineering has become science's quest to improve nature's products. In principle, almost any desirable trait found in nature can be transferred into any chosen organism.

Genetic engineering is the general process of transferring DNA from one organism, usually a plant or animal, to another. The result of genetic engineering is a transgenic plant or animal, also known as a genetically modified organism. Transgenic organisms contain DNA from a different kind of plant or animal. The transferring of genes can be performed in various ways. Cells can be directly injected with DNA, which is literally shooting DNA-covered particles from a special gun into cells. Another way of transferring genes is by inserting DNA into specially modified bacteria or viruses that carry it into the desired cells.

## What Are Genetically Modified Products?

A genetically modified product is a product that is produced by a genetically engineered organism. Genetically engineered products are used in pharmaceuticals, transgenic plants and animals, and gene therapy. Human insulin used by diabetics and animal drugs used as growth hormones, such as bovine or porcine somatropin, are examples of pharmaceuticals produced using genetically modified organisms. These products are produced using bacteria that have received the appropriate human, cow, or pig DNA. Transgenic plants that can tolerate herbicides, resist insects or viruses, or produce modified fruit or flowers are currently being grown and tested, and in some cases being used or marketed. Examples of these transgenic plants are corn plants that produce an insecticidal protein that resists European corn borers and tomatoes that can ripen longer on the vine before shipping. Transgenic animals are being designed to help researchers diagnose and treat human diseases. By testing transgenic mammals that produce important pharmaceuticals in their milk, scientists can determine if products such as insulin and growth hormones may be available from milk of transgenic cows, sheep, or goats. Gene therapy is a new treatment that uses genetically modified products. When a gene is missing or not functioning correctly, it can be replaced with a correct gene by using gene therapy. The first successful gene therapy occurred in 1990 and was used to treat an immune system defect called adenosine deaminase (ADA) deficiency in children. Blood cells with normal ADA genes were injected into patients where they produced normal cells used to improve the patients' immune systems. Currently, gene therapy clinical trials are being used to treat diseases such as malignant brain tumors, cystic fibrosis, and HIV. The entrance of health-related genetically engineered products into the marketplace usually has not generated the same level of public controversy that surrounds proposed introduction of genetically engineered foods.

## How Are Genetically Modified Organisms Regulated?

Three different government agencies watch out for consumers' interest by regulating genetically modified organisms (GMOs). The U.S. Department of Agriculture (USDA), Food and Drug Administration (FDA), and the Environmental Protection Agency (EPA) are responsible for telling consumers which products made using GMOs are safe. State governments also have a role in regulating GMOs.

The following information was taken from the respective agency websites. For the most current information, check the specific agency website or the U.S. Regulatory Agency's Unified Biotechnology website (http://usbiotechreg.nbii.gov/lawsregsguidance_agency.asp).

## U.S. Laws and Regulations

### U.S. Department of Agriculture

Plant Protection Act (PPA); (Title IV of the Agriculture Protection Act) http://www.aphis.usda.gov/ppq/plantact
Regulations issued under Plant Protection Act-7 *CFR* Part 340: Introduction of Organisms and Products Altered or Produced Through Genetic Engineering Which are Plant Pests or Which There is Reason to Believe are Plant Pests-http://www.aphis.usda.gov/brs/7cfr340.html.[1]

On June 14, 2002, Animal and Plant Health Inspection Service (APHIS) created the Biotechnology Regulatory Services (BRS) to place increased emphasis on our regulatory responsibilities for biotechnology. However, while BRS was established fairly recently, APHIS has a long history of regulating agricultural biotechnology products, overseeing the safe conduct of more than 10,000 field tests of genetically engineered crops, and the deregulation, or removal from government oversight, of more than 60 products.[2]

While biotechnology holds enormous potential for reducing herbicide use, increasing crop health and production, and manufacturing medicines and industrial products, the challenges posed by biotechnology highlight the importance of the regulation of this technology. APHIS BRS responsibility is to ensure a dynamic, robust regulatory system based on science and risk that ensures safe field testing and product development in the United States and is mindful of the global implications of our work.[2]

Companies and organizations who wish to field test a genetically engineered crop must obtain USDA permission through a permit. Companies must submit all plans for field testing for review by regulatory scientists who evaluate the risks of the test and the protocols to be employed. USDA will approve the plan if the proposed test conditions appear adequate to confine the regulated article within the field test site. To ensure compliance to the permitting conditions, field test sites are inspected and records are audited.[3]

Depending on the nature of the genetically engineered crop, an applicant files either a notification or a permit application. In general, most of the plants are field tested under the notification procedure, a more streamlined approval process that is used only for familiar crops and traits considered to be low risk. Permitting is used for field tests of plants that have an elevated risk, such as plants producing pharmaceutical or industrial compounds.

Notifications are less restrictive than permits. They are used for low-risk crops—weeds and traits with higher risk are excluded. BRS reviews the application for completeness, taking up to 30 days to process the application. Performance standards have been established, and an applicant must comply with these for movement, planting, growing, harvesting, and isolation.

Permits are more restrictive than notifications. They are used for crops with elevated risk, including pharmaceutical and industrial products and plant pathogens. Scientists review the conditions and confinement of the GMOs. The application process requires up to 120 days to complete. BRS authorizes procedures for field production and isolation. All organisms and all traits are eligible.[3]

Comprehensive permits reduce the number of notifications for applicants, APHIS, and states. Genes can be listed in a new alternative style that allows any combination of genes of interest, regulatory genes, and marker genes listed to be tested. Sites and movements can be quickly and easily added for states and genes in the issued permits. Notifications can still be used for entries not anticipated at the time of submission.

Applicants must adhere to the conditions of the permit or notification. Applicants testing products are required to report infractions, accidents, and unusual occurrences. APHIS officers or BRS staff may inspect field sites and check records. Some infractions, investigation, and deliberation may result in civil fines and compensation for damage or remediation. All field tests may be inspected and audited by USDA APHIS personnel. In the period 1990–2001, only about 2 percent of field tests received a "compliance infraction" action by the agency.

To remove a regulated article from BRS oversight through field testing, an applicant must provide a petition with complete product information and environmental safety data. BRS reviews the data for completeness and accuracy and has 180 days to make the decision to deregulate, though the process may require more time to complete. An environmental assessment is written to analyze the possible impacts, and public comment is sought. A petition is approved when a finding of no significant impact is reached. Those petitions that have resulted in deregulation, or are in the process of deregulation, are listed at http://www.aphis.usda.gov/brs/not_reg.html.[3]

Developers of new genetically engineered products can petition APHIS for a determination of nonregulated status of these products. Deregulated articles are no longer subject to APHIS oversight and can be used in food, feed, and in breeding programs in the same way as conventional products.

The Coordinated Framework for Regulation of Biotechnology, 51 *Fed. Reg.* 23,302,(1986) prescribes how products of U.S. biotechnology be reviewed and what federal agencies will review them. Separate but coordinated reviews are made by the USDA, EPA, and FDA, depending upon the gene and host crop.

### U.S. Environmental Protection Agency

Federal Insecticide, Fungicide and Rodenticide Act (FIFRA; 7 *USC* Chapter 6). http://www.epa.gov/epahome/laws.htm

Federal Food, Drug, and Cosmetic Act (FFDCA; 21 *USC* Chapter 9). http://www.epa.gov/epahome/laws.htm

Regulations issued under FIFRA and FFDCA 40 *CFR* Parts 152 and 174: Pesticide Registration and Classification Procedures http://www.epa.gov/epahome/rules.html#codified

40 *CFR* Part 172: Experimental Use Permits. http://www.epa.gov/epahome/rules.html#codified

Toxic Substances Control Act (TSCA); 15 *USC* Chapter 53). http://www.epa.gov/epahome/laws.htm

Regulations issued under TSCA: 40 *CFR* Part 725: Reporting Requirements and Review Processes for Microorganisms http://www.access.gpo.gov/nara/cfr/cfr-retrieve.html#page1

The regulation under which the TSCA Biotechnology Program functions is titled Microbial Products of Biotechnology; Final Regulation Under the Toxic Substances Control Act, promulgated in the *Federal Register* on April 11, 1997. This rule was developed under TSCA Section 5, which authorizes the agency to, among other things, review new chemicals before they are introduced into commerce. Under a 1986 intergovernmental policy statement, intergeneric microorganisms (microorganisms created to contain genetic material from organisms in more than one taxonomic genera) are considered new chemicals under TSCA Section 5. The biotechnology rule sets forth the manner in which the agency reviews and regulates the use of intergeneric microorganisms in commerce or commercial research.[4]

## Final Regulations Under the Toxic Substances Control Act Summary

The EPA has published final rules that fully implement its screening program for new microorganisms under Section 5 of the TSCA. These regulations tailor to microorganisms the screening program that had been in place since 1986 for microbial products of biotechnology. They established a separate part in the *Code of Federal Regulations* for microbial products of biotechnology subject to TSCA, 40 *C.F.R.* Part 725; created a number of exemptions; and codified the EPA's approach to research and development (R&D) for microbial products of biotechnology. These rules provided significant regulatory relief to those who wished to use certain products of microbial biotechnology. At the same time, these rules were designed to ensure that EPA could adequately identify and regulate risk associated with microbial products of biotechnology without unnecessarily hampering this important new industry.

Microorganisms subject to this rule are "new" microorganisms used commercially for such purposes as production of industrial enzymes and other specialty chemicals; agricultural practices (e.g., biofertilizers); and breakdown of chemical pollutants in the environment. These rules continue the interpretation of *new* microorganism first put forth by EPA in 1986. New microorganisms are those microorganisms formed by combining genetic material from organisms in different genera (intergeneric). A genus (pl. genera) is a level in a classification system based on the relatedness of organisms. The EPA believes that intergeneric microorganisms have a sufficiently high likelihood of expressing new traits or new combinations of traits to be termed *new* and warrant review. Microorganisms that are not intergeneric would not be new and thus would not be subject to reporting under Section 5 of TSCA.

These regulations created a reporting vehicle specifically designed for microorganisms, the Microbial Commercial Activity Notice (MCAN). Persons who intend to use intergeneric microorganisms for commercial purposes in the United States must submit an MCAN to the EPA at least 90 days before such use. The EPA has 90 days to review the submission in order to determine whether the intergeneric microorganism may present an unreasonable risk to human health or the environment.[5]

The rules also address intergeneric microorganisms used in R&D for commercial purposes and created a vehicle for reporting on testing of new microorganisms in the environment, a TSCA Experimental Release Application (TERA). A TERA should be submitted to the EPA at least 60 days prior to initiating such field trials. The TERA is designed, in recognition of the needs of researchers, to provide a high measure of flexibility and a shorter review period (60 days).[5] R&D for commercial purposes is that activity that is funded directly, in whole or in part, by a commercial entity regardless of who is actually conducting the research or that will obtain for the researcher an immediate or eventual commercial advantage.

Certain intergeneric microorganisms are exempt from the requirement to submit a MCAN if the manufacturer meets criteria defining eligible microorganisms and specified use conditions. This exemption is most applicable to the manufacture of specialty and commodity chemicals, particularly industrial enzymes.[5]

Intergeneric microorganisms used for R&D in contained structures are exempt from EPA reporting requirements if researchers maintain records demonstrating eligibility. Researchers are exempt from this recordkeeping requirement when the researcher or institution is in mandatory compliance with the National Institutes of Health (NIH) Guidelines for Research Involving Recombinant DNA Molecules. Those researchers voluntarily following the NIH guidelines can, by documenting their use of the guidelines, satisfy the EPA's requirements for testing in contained structures. Alternatively, researchers can take the exemption by documenting that they meet eligibility criteria laid out by the EPA in these rule makings.

Certain intergeneric microorganisms in R&D field testing are also exempt. Testing on 10 acres or less involving *Bradyrhizobium japonicum* and *Rhizobium meliloti* is exempt when certain exemption criteria specified by these rules are met.[5]

### U.S. Food and Drug Administration

Federal Food, Drug, and Cosmetic Act (FFDCA); 21 *USC* Chapter 9). http://www.epa.gov/epahome/laws.htm
Regulations and Guidance issued under FFDCA: Statement of Policy: Foods Derived From New Plant Varieties (1992). http://www.cfsan.fda.gov/~acrobat/fr920529.pdf
Consultation Procedures under FDA's 1992 Statement of Policy (1997). http://www.cfsan.fda.gov/~lrd/consulpr.html

Under this policy, foods such as fruits, vegetables, grains and their by-products, derived from plant varieties developed by the new methods of genetic modification are regulated within the existing framework of the act. The FDA's implementing regulations and current practice use an approach identical in principle to that applied to foods developed by traditional plant breeding. The regulatory status of a food, irrespective of this method by which it is developed, is dependent upon objective characteristics of the food and the intended use of the food (or its components). The method by which food is produced or developed may in some cases help to understand the safety or nutritional characteristics of the finished food. However, the key factors in reviewing safety concerns should be the characteristics of the food product rather than the fact that the new methods are used.

The safety of a food is regulated primarily under the FDA's postmarket authority of section 402(a)(1) of the act (21 *U.S.C.* 342(a)(1)). Unintended occurrences of unsafe levels of toxicants in food are regulated under this section. Substances that are expected to become components of food as a result of genetic modification of a plant and whose composition is such or has been altered such that the substance is not generally recognized as safe (GRAS) or otherwise exempt are subject to regulation as food additives under section 409 of the act (21 *U.S.C.* 348). Under the act, substances that are food additives may be used in food only in accordance with an authorizing regulation.

In most cases, the substances expected to become components of food as a result of genetic modification of a plant will be the same as or substantially similar to substances commonly found in food, such as proteins, fats and oils, and carbohydrates. The FDA has determined that such substances should be subject to regulation under section 409 of the act in those cases when the objective characteristics of the substance raise questions of safety sufficient to warrant formal premarket review and approval by the FDA. The objective characteristics that will trigger regulation of substances as food additives are described in the guidance section.

The guidance section also describes scientific considerations that are important in evaluating the safety and nutritional value of foods for consumption by humans or animals, regardless of whether the food is regulated under section 402(a)(1) or section 409 of the act.

## Guidance Section

This section describes many of the scientific considerations for evaluating the safety and nutritional aspects of food from new plant varieties derived by traditional methods (such as hybridization or mutagenesis), tissue culture methods (such as somaclonal variation and protoplast fusion), and recombinant DNA methods. Although some of the safety considerations are specific to individual technologies, many safety considerations are similar regardless of the technology used. The FDA expects plant breeders to adhere to currently accepted scientific standards of practice within each technology. This guidance section is based on existing practices followed by the traditional plant breeders to assess the safety and nutritional value of new plant varieties and is not intended to alter these long-established practices or to create new regulatory obligations for them.

## Examples of a Genetically Engineered Food

### The Flavr Savr Tomato

The Flavr Savr tomato manufactured by Calgene, Inc. was the first whole food product produced by biotechnology to be evaluated by the FDA. This tomato was engineered to stay firm after harvest, allowing it to stay on the vine longer and still be able to make it to market without getting crushed. This was accomplished by inserting DNA from *Agrobacterium tumefaciens, Escherichia coli,* cauliflower mosaic virus, and tomato. The net result of the new DNA was the suppression of the polygalacturanase enzyme, which is associated with the breakdown of pectin. This breakdown is what leads to the softening of the tomato. A selectable marker gene that confers kanamycin (an antibiotic) resistance was used so that Flavr Savr tomatoes could be identified.

FDA evaluated the Flavr Savr to determine if it was as safe as other currently consumed tomatoes. They did this by looking at the genetic material inserted in the plant, the chemical composition of the tomato, and the safety of the genetic marker used. This marker was the main concern because it conveys resistance to the antibiotic kanamycin. Calgene, Inc. determined that the resulting protein could be easily broken down in human stomachs and that it would not be able to effect the therapeutic efficacy of orally administered antibiotics.

## U.S. Inspection System for Biotech[6]

APHIS's quick response to finding unapproved bioengineered corn plants in field test sites in Iowa and Nebraska in 2002 demonstrated the effectiveness of the agency's routine inspection system. In that case, APHIS took action to keep the product out of the food supply and is requiring the company that managed the field tests, ProdiGene, Inc., to follow additional biotechnology field test requirements. The agency also issued the company a severe monetary penalty. In addition, APHIS applied new requirements for field tests of pharmaceutical crops—or pharma crops—grown for vaccines and other medical uses. Included are requirements for mandatory record keeping, increased buffer zones around test areas, equipment dedicated only to the field tests, and mandatory training for field personnel. The agency also increased the frequency of its inspections.

## References

1. National Biological Information Infrastructure (NBII). United States Regulatory Agencies Unified Biotechnology Website. Online http://usbiotechreg.nbii.gov/lawsregsguidance_agency.asp (September 2004).
2. USDA. 2002. Biotechnology Regulatory Services (BRS) Website. Online http://www.aphis.usda.gov/brs/welcome.html (September 2004).
3. USDA. Regulatory Activities of BRS Website. Online http://www.aphis.usda.gov/brs/regulatory_activities.html (September 2004).
4. EPA. 2004. Biotechnology Program Under Toxic Substances Control Act (TSCA). Online http://www.epa.gov/opptintr/biotech/biorule.htm (September 2004).
5. EPA. 2004. Microbial Products of Biotechnology: Final Regulations Under the Toxic Substances Control Act Summary (Fact Sheet). Online http://www.epa.gov/biotech_rule/fs_001.htm.
6. APHIS. 2003. U.S. Inspection System for Biotech Effective, Official Says. Online http://usinfo.state.gov/ei/Archive/2004/Jan/28-787641.html (September 2004).

# Index

**Note: Tables are indicated with *t* following page locator.**

U



V



W



Y



Z